W9-AVQ-881

THE FIRE THAT WILL NOT DIE

Michele McBride

An ETC Publication

Library of Congress Cataloging in Publication Data

McBride, Michele, 1945-
The fire that will not die.

1. Burns and scalds—Biography.
2. McBride, Michele, 1945-
3. Chicago—Fire, 1958.
4. Our Lady of the Angels (School)—Fire, 1958. I. Title

RD96.4.M28 362.1'9'7110926 [B] 78-31781
ISBN 0-88280-066-3

Published by ETC Publications
Palm Springs
California, 92263

Printed in the United States of America.

Dedication

Like the mystic bird, the Phoenix, I have met the challenge of fire, I have grown. In loving memory I dedicate this book to the children who did not have the opportunity to meet the many challenges of life, and whose experiences in life were all too short. I write in hopes that they are called to memory for they were once children of this earth and deserve to be remembered.

The following is a list of children who died as a result of the fire which swept Our Lady of the Angels School, Chicago, Illinois, December 1, 1958.

Michele Altobell
Robert Anglim
Karen Baroni
David Biscan
Richard Bobrowicz
Beverly Burda
Helen Buziak
Peter Cangelosi
George Cannella
Kathleen Carr
Margaret Chambers
Aurelius Chiapetta
JoAnne Chiappetta
Joan Chrzas
Bernice Cichocki
Rosalie Ciminello
Roseanna Ciochon
JoAnn Ciolino
Milicent Corsiglia
Karen Culp
Maria DeGiulio
Nancy De Santo
Patricia Drzymala
Lawrence J. Dunn

William Edington
Mary Ann Fanale
Lucile Filipponio
Nancy Finnigan
Ronald Fox
Janet Gasteier
Carol Ann Gazzola
Lawrence Grasso
Frances Guzaldo
Kathleen Hagerty
Richard Hardy
Karen Hobik
Barbara Hosking
Victor Jacobellis
John Jajkowski
Angeline Kalinowski
Richard Kompanowski
Kenneth Kompanowski
Diane Karwacki
Joseph King
Margaret Kucan
Patricia Kuzma
Annette LaMantia
Rose Anne La Placa
Joseph Maffiola
Raymond Makowski
Linda Malinski
John Manganello
John Mele
Joseph Modica
James Moravec
Mary Ellen Moretti
Charles Neubert
Lorraine Nieri
Janet Olechowski
Yvonne Pacini
Antoinette Patrasso
Eileen Pawlik
Carolyn Perry
Elaine Pesoli
Mary Ellen Pettenon
Edward Pikinski

Nancy Pilas
Frank Piscopo
James Profita
James Ragona
Roger Ramlow
Marilyn Reeb
Nancy Riche
Margaret Sansonetti
Diane Santangelo
Joanne Sarno
William Sarno
Antoinette Secco
Kurt Schutt
James Sickels
Paul Silvio
Susan Smaldone
Nancy Smid
Linda Stabile
Mark Stachura
Mary Tamburrino
Philip Tampone
Valerie Thoma
John Trotta
Mary Virgilio
Christine Vitacco
Wayne Wisz

Acknowledgements

I wish to thank LaVelle Frawley, Kathleen Goneno, Jacquelyn Hershewe, Barbara Melton, and Ralph Van Atta for their encouragement to write this book. Without their constant support I could not have finished it.

The names in this book have been changed so as not to cause more pain for families associated with the disaster.

Contents

INTRODUCTION

This is a story about a disaster, the burning of Our Lady of the Angels school in Chicago on December 1, 1958, in which ninety-two children and three nuns perished. My story is a first-hand account, because I am an Our Lady of the Angels survivor. I am one of the "victims" who somehow managed to survive a massive body burn. I was thirteen years old when fire struck my school, and since that time I have been forced to struggle against continuous pain and heartache. My story describes what I have endured as a post-burn person who had to face not only the ordeal of healing a burned body, but the anguish of learning to accept life again.

I wish I could say that it was bravery, superhuman courage, some inner, heaven-sent strength that sustained me through the agony, but I cannot. It was anger, raging anger that made me survive. I was angry at the lack of authority in my classroom when the fire broke out. I was angry because the firemen's ladders fell short of the classroom windows, because I lost the skin of my birthright, because I had to endure ravages of pain that I had thought were reserved for those condemned to the torments of Hell. I was angry for having lived, and I was angry at those who did die and left me behind. I was angry at being treated like a child after I had witnessed millions of years of burning all condensed into a single moment. Hellfire, the witches of Salem, the melting skin of Hiroshima—I saw them all, and yet I never left my classroom. All the horrors of the world were presented to me in one brief second and made me realize I am mortal, I shall die.

In writing this book I have had to come to terms with many unspoken fears and ideas. It has been difficult, because I was taught very early to think that the fire was the mystical act of a lonely God, and that it should not be talked about or questioned. In my community it was always felt that bringing up memories of the fire would just add new grief to the old. I disagree. I could not recover from the fire until I had learned how to mourn, not only for my dead friends but for my skin as well. I think that discussing a disaster and remembering the dead can help to heal wounds and resolve anguish in any stricken community.

Through the fire and its aftermath, I learned that disaster does not breed those strong, jolly, humble heroes that we read about in newspapers and books. Real survivors experience anger, panic, jealousy, guilt, self-doubt--all those feelings people never like to talk about, but which are as important and as powerful as bravery, kindness, and love. My story is about those neglected feelings, but it is also about humor, that lifesaver in times of stress. I think a quick wit is the most resilient quality a human being can possess, and I cannot help suspecting that people who say humor is a cover-up to hostility, just don't have a sense of humor.

I was often embarrassed by all my feelings because I did not understand it was necessary to let all emotions surface. When people told me how brave I was, I always resented it. Bravery is a matter of choice, and I never volunteered for my position in the disaster. I simply acquired it, for the most part, because I misjudged the danger of fire. I did not know how fast fire travels or that hot smoke can kill. I thought that walls were safeguards that neither fire nor smoke could penetrate. I thought that I could safely wait for the firemen to come. I did realize that time was precious, and I was shocked when the teacher told us to say the Rosary, the longest prayer in the Catholic Church. It was such an inappropriate command, and yet what else could she have said? We were trapped, and some of my classmates already knew it. It was the children who felt the real impact of our disastrous situation that began the work of survival.

It has taken me five years to write and document this account of the holocaust. Much of that time was spent in waiting for a happy ending that never materialized. I want to establish a national rehabilitation program for post burn patients, but no such program exists, and I have not been able to create one. I am disappointed--or perhaps I should say I am angry. Anger has brought me through other difficult times, and I am sure it will not fail me now.

Being a fire victim has caused me much grief, but it has given me unique opportunities as well. It was very difficult to obtain official information about the Our Lady of the Angels fire. The Chicago Fire Department and the Archdiocese of Chicago would not let me examine their records; they said that their files were closed to the public. However, it was the public who provided me with the information I needed by sharing their stories and feelings with me. It was a heart-warming experience to converse with the classmates, parents, and rescue workers who welcomed me because I was a fire victim. They gave me the courage to continue my research, and I thank them.

Michele McBride
Chicago, Illinois

Part One

Fire! Fire!

CHAPTER I

December 1, 1958 was the first Monday of the month and the first day of Advent, so I went to Mass as I would be going every day of that holy season leading to Christmas. There was nothing special about that day; even my horoscope did not indicate anything extraordinary was to take place later that afternoon.

I was particularly happy that morning. I was well enough to be returning to school after an absence of a few days due to a painful attack of bursitis, and I would see my friend Helen again. I knew she would be in school early because I had already seen her in church that morning. I always went to the seven o'clock Mass and she attended the seven-thirty Mass, so we waved as we passed each other.

Helen was treated somewhat special. She was given the privilege of eating breakfast in the classroom because she lived too far to go home for breakfast after Mass. I thought it must be nice to carry your breakfast to school because you got to eat doughnuts. Helen always had those tiny sugar-covered doughnuts and she shared them with me. They were the kind that melted in your mouth almost before you had a chance to bite into them, and a fine layer of the confectioners sugar always drifted down onto the lap of our blue uniform skirts.

I remember that morning. There was a brisk wind, and it was a clear day.

FIRE! FIRE!

I took my favorite shortcut down the alley. There were dried clumps of weeds twisted in and around the fenceposts, and large balls of stale weeds frolicked around and around as the breeze blew them up and down the alley. At that time of the year the alley looked like a charcoal pencil sketch with all the varying somber hues, the darkness of the black tar pavement contrasting with the gray and white fences.

I could see several shades of slate gray and beige on the different cement aprons behind the garages in the backyards. There was a thick ebony-colored crack running almost the entire length of the alley. This crack marked an invisible dividing line so everyone knew just how far his property rights extended into the alley. The tiny diamond shapes in the chain link fences added a dotted pattern to the perpendicular slate of the wooden fences.

It wasn't a gray day, just a gray alley, awaiting the first snowfall of the season. The alley was still cluttered with reminders of the passing autumn, with weeds and leaves blowing around, playing hide and seek with each other.

I happily scuffed aside some dried leaves as I hurried along and thought how nice it was to be returning to school after missing so much time because of my painful attack of bursitis.

Right at the corner where I turned out of the alley I could see the school. Our Lady of the Angels Church towered over the school and the neighborhood of tiny, snug bungalows and apartment buildings. The school consisted of two large rectangular buildings connected by a third annex building forming a large U-shape. The orignial building (south wing) built in 1904 housed a school on the first floor and a convent on the second floor. In 1910 a combination church-school was built on North Avers Avenue. It was connected to the original building by an annex. All the construction of each building was two story brick, wood joisted. In 1939 the cornerstone was laid for a new church and rectory on adjoining property. The church-school (north wing) was converted to all classrooms with a chapel in the basement. In 1951 the remodeling of the interior of the school was completed. The maroon brick, trimmed with the mustard seed yellow of the windows, always reminded me of a huge package ready to be mailed. The straight little rows of window trim running throughout the building looked like string on this package.

It was early when I got to school that day and Helen was already in the room eating breakfast. We were both there before the school bell rang because we had volunteered to dust the room. I think in elementary school, I dusted more than I studied. Even today I can recall the smallest details of the windows and moldings because I dusted thousands of feet of window sills for eight years. I can still see the high enamel sheen of the brown painted window sills and the dull mustard seed yellow trim of the outside.

There was a little hole on the top of the window sash for opening each window. I thought it was the cleverest thing to be able to open the windows not just from the bottom, but the top, too. It was always a boy's job to open and close the top windows with the window pole. I do not know why this was a male dominated assignment, because it never seemed to me to be that difficult to place a small metal hook into the hole at the top of the windows. I could never quite figure out just what the qualifications were to be chosen to close the windows. Tall boys were called, short boys were called, well-behaved boys were chosen, mischievous boys were chosen, but never the girls. Unless, of course, no boys were available. Whoever was lucky enough to have charge of this much sought after task had to take care never to knock down one of the forest-green window shades. No one would ever adjust the windows without permission from the teacher in charge.

Helen ate her breakfast and we talked of the movie on the late show the night before. While I dusted the window sills I looked out into the courtyard and thought how drab it was to be in the north wing section of the school. We were the "Eighth Graders," supposedly the upper classmen who had made the big time, and had the worst seats in the house. We had nothing to see but an ugly little courtyard with a tall, imposing iron-spiked fence keeping out trespassers. No one was ever allowed to go into that yard except the boys who beat the chalk dust out of the erasers by banging them against the building. Directly across the courtyard from our windows were other windows looking into another classroom filled with students just like us.

I did not like being in one of the rooms on the inside of the "U" shaped school building and would have preferred being in one of the outside rooms facing the tall green trees. Trees were fun to watch and I enjoyed seeing the branches swaying when the breeze tickled them. From our window, there were no trees. There was nothing to see but the dreary brick courtyard and the only thing to break up this mortar and brick view was a small porch located in the corner right under the window where I sat. It was painted the regulation mustard seed yellow and brown.

As I dusted that morning, I could not reach behind the radiator because I still had my arm in a sling due to the bursitis. I also had oil of wintergreen rubbed on my shoulders and on my left arm and smelled like a walking medicine cabinet—I hoped no one would notice that I was the one who smelled like a cross between a gum drop and a bottle of vinegar.

Helen finished doing all the places I could not reach. This was the last time I would see her being herself. Before nightfall we would have left a child's world to experience a world where cries of demons were mere chit chat.

Soon the school bell rang and rows of children streamed into the room. Helen sat in the front of the room and I in the back, but we passed notes during the day to keep in touch. One of the special aspects of our friendship was our being the only two girls allowed to go downtown to shop for the nuns. They frequently sent us shopping to buy supplies for bulletin boards, and a certain type of long, thin, white headed pin used to fasten and hold a part of their habits. This was like sharing some mystery of the Catholic church, knowing which section in Marshall Field's department store carried these special pins.

During religion class that morning our teacher made mention

that Helen and I and one boy were the only three who went to Mass the first day of Advent season. I am sure we were quite pleased with our efforts and everyone else thought we were big pains.

I remember that the morning dragged as we followed the usual schedule of subjects, but at lunchtime Sister told us that a new system for dismissal had been devised. Instead of the entire class walking to the front of the church, we marched out and broke ranks on the corner nearest the direction of our homes.

Procession lines, marching, and fire drills were as important in our Catholic school as arithmetic and spelling. I think that some of the nuns were frustrated choreographers with all the practicing we did for such occasions. Fire drills were important and there was a regimented, prescribed method for evacuations. Again and again the nuns told the students, "Stand up quickly, leave everything at your desk, line up and march out in single file, arms held down at your sides, no talking." Above all, there was to be no grabbing for the hand rails on the stair, as this might waste time and we were to get out of the building as fast as possible.

We were drilled. Twelve hundred students of Our Lady of the Angels could march out in just three minutes using all the hallways and stairwells. But the drills were all in vain because the day the fire came, it intruded on the prescribed escape method and raged through the stairwells, blocking off the hallways and leaving no escape routes.

That noon Forty Hour Devotion was scheduled in the church because it was our parish's turn to display the Blessed Sacrament. Some of us children visited the church before lining up to go back into school after lunch. We then marched inside in an orderly manner, as we had been trained, and hung our coats on the coathooks which accommodated about three hundred students' coats, or wraps. I always felt like a chocolate bar when the nuns said "Go get your wraps," it was like being a Hershey bar leaving the chocolate factory to get wrappers.

Funny the language nuns used: parlors were living rooms,

rubbers were boots, and fire was "pray."

Everyone had to have a cloth loop sewn into coats, jackets and sweater collars in order to keep them securely hanging on hooks so they would not litter the hallways in the event of fire. Strange how much confidence we put into small details—details that do not mean a whole lot when fire strikes.

Also located in the hallway was a fireman's pick that looked like a red harpoon. It was silly to have harpoons in the hallways, since we did not have any whales enrolled in school, but in this manner we were protected from falling coats and whales.

In the afternoon I was busy with my arithmetic paper and it was difficult because I had the use of only one arm. I still kept my left arm drawn close to my body so the weight of the shoulder was confined to the sling around my neck. Most of the class finished their papers and the row captains were collecting them. My captain already sailed by me which was usual. He was a prompt person, I, a procrastinator. I would have to tap the shoulder of the girl who sat in front of me and pass my paper up to the teacher's desk via the touch system. It must have been about twenty-five to three with only ten minutes left of school.

Suddenly, Jeff, a boy sitting about five rows across from me, jumped from his desk yelling something which I could not understand at first. He reached for the rear door and barely touched it when he yelled, "Jesus Christ, it's hot!"

The class was shocked that he would swear in front of the nun who ignored his swearing and ran to the front door. When she could not open it with her bare hands, Sister took a corner of her habit and managed to force the door open. Big blasts of black smoke rushed into the room. The door slammed shut; the force of the smoke was so great. Some children helped the teacher stuff soft cover workbooks around the front door, but this did not keep the smoke from sweeping into the room. Children screamed, climbed over desks and many ran to the rear door, but it was too hot to push open.

There was a stampede of frightened students toward the

windows. Rows of desks were pushed toward the doorways. The desks near the windows were intact for the moment. The sound of the scraping, clapping, metal and wooden framed desks crashing into one another was startling. Students began throwing books out of the windows shouting, "Fire, fire. The school's on fire."

I don't think anyone grasped the immediate danger and Sister must not have either because she told the class to recite the Rosary, the longest prayer any of us knew.

There was a futile attempt to start the Rosary by some students who started to recite the solemn prayer, "The Apostles Creed," but their voices were drowned by the screams, "Fire, fire. The school's on fire."

I kept thinking the walls would protect us from the fire and that the firemen would come soon. What frightened me more than the fire was the panic in the room. Seeing classmates change into struggling monsters, begging for air was worse than the danger of fire. I stood at my desk expecting to hear someone call out the order to line up and march out single file. I expected control over the rioting group. I was shocked at the lack of order in a room where orderliness was the rule, and I could not comprehend our peril. I was ignorant of the dimensions of fire. Instead of reaching out, I was withdrawing and I kept reassuring myself that soon everything would be just fine again.

"The firemen will come," I said over and over. "The firemen will come."

In desperation we started throwing books, our shoes, everything we could out the windows to attract attention. Children stood on the window sills and tore the green shades from their brackets and tossed them down into the room. Someone grabbed the long window pole and started to smash the windows. Fragments of glass floated down through the air leaving big jagged pieces in the window frames, but the holes didn't let in more air, only more thick smoke out.

I was horrified as I watched the window pole banging and crashing against the glass. I looked up through the jagged pieces of glass and saw, for the last time, the blue serene sky as it was being besmirched by this greasy, black smoke pouring over our heads.

I was pinned against a window sill by a crowd of students; someone was standing on my neck crushing my face into the window sash. I pushed away from there, it was too dangerous. I was choking.

I stood free from the crowd when a small girl started to scream. I slapped her face as hard as I could and I was startled by my own reaction and surprised she did not strike me back. Two boys carried her to a window.

Just then Teddy, a husky mischievous fellow who always had a mind of his own, pushed his classmates out of the way and screamed, "I'm getting out of here," and he did. For a second he perched on the window frame and then jumped out of the corner window as other trapped students screamed. Teddy landed on the roof of that silly, little, useless porch right beneath the window, his hugeness making the tar paper roof island look even smaller.

We breathlessly watched him grab on to the rain pipe, which split and swung out from the building. Teddy dangled in midair for a brief second and then jumped about ten feet more to the ground. He stood safely on the cement courtyard and shouted up that he was going to get help.

I was greatly impressed by his fast action. Soon other children followed his example, jumping down about fourteen feet to the little porch roof and easing themselves down the broken rain pipe, screaming all the way. It seemed as though we all would be safe in a few minutes, but fire does not wait calmly for minutes to tick away while aid is being summoned, It knows no measurement of time and never waits.

"Teddy will go get the firemen. The firemen will come and take us safely out of this room," I kept reassuring myself.

I was again pushed up to the first line of students at the windows and saw a girl struggling against the crowd. As she tried to jump out, her blue skirt caught on the pencil sharpener which was fastened to the window sill. Boys and girls screamed at her to get out of the way, and were grabbing and pulling her skirt free. Another boy climbed over her and jumped down to the little porch roof, and fell to the ground. I drew back because I simply could not manage the jump down that far and I was fearful I would be pushed out of the window by the panicky crowd.

"The firemen will come," I thought to myself.

CHAPTER II

Students across the courtyard began waving back at us as though entering into a sport. Someone over there started to draw the window shades closed, as if we were making too much noise, but they caught on to the dangerous plight we were in when they saw our teacher frantically waving her arms.

The fire bell finally rang and we all felt reassured when the familiar foghorn bellowed loudly throughout the entire school building. We watched children in the opposite wing marching out by the established fire drill procedure, wishing we could do the same.

"The firemen will come," I whimpered to myself.

I stepped back to my desk and reached in for a handkerchief. Believing that tomorrow we would all be back to collect our personal belongings, I decided to leave my blue and white wallet inside my desk. I never dreamed of the destruction that was ahead of me.

The girl who sat in the desk in front of me was crying and throwing her books in the air. The windows I had trouble dusting that morning were shrinking in size with the huge pack of humanity clammering to get a breath of air. I covered my mouth with my hanky and decided to push my way to the front of the room, but a girl was standing in the way hysterically crying, "I don't want to die." I just slapped her in the face and said, "Shut up. The firemen will come."

The steady blaring racket of the fire bell, so loud and clear, was a steady reminder that the firemen would come.

I saw little girls dressed in navy blue uniform skirts start beating on their legs because the bare skin was starting to smart from the heat. I watched in horror as children stopped struggling for air and started laughing. It was not a friendly, joking laugh, but the snarling bray of a hyena. Their heads were thrust back, their mouths wide open and gagging, as spit streamed down their fear stricken faces.

I could see down into the courtyard where the iron fence stood so ominously and watched men dressed in overalls gather outside beating on the metal. These men were old and worn, the fence was old and weather beaten, but indestructible. These were the grandpas of the parish who daily sauntered around the neighborhood visiting each other or tended to their small gardens. Many of them had come from Europe and had undergone the terrors of war, but not one of them was prepared for the horror they were to see that day—children on fire dropping from high windows and crashing to the ground.

These elderly men stood with their arthritic hands, twisted and toughened from years of hard factory work, gripping the bars of the iron blockade. Frantically they rubbed their clenched fists up and down the iron bars as though trying to wear the metal away until splinters of rust broke the skin and their hands bled. The men hurled their bodies against the mighty fortress in a frenzy of despair, but the fence did not yield. It stayed. Black bars towered over their heads implacably holding them back, and the men wept when they could not reach the children who were hanging out of the windows on the second floor.

Someone brought a sledge hammer and the old men took turns swinging this full force against the iron fence, but the only result was the deafening ringing sound of metal on metal. The solemn clangor rang and echoed in the courtyard, but the fence did not move.

A ladder was thrown over the fence and the men who had fought back the smoke and crawled into the courtyard through the doorway of the little porch, raised it up towards the classroom window. It was too short. Another ladder was thrown over the iron fence. The ladders tied together, almost reached a girl who swayed down and barely touched the top of the ladder with her feet. The ladder collapsed and the child smashed to the ground.

Crowds of people were gathering outside the iron fence. Boys leaned out of the windows and called for their mothers. Mothers were calling for their sons. Women in housedresses were running to and fro behind the bars screaming, "Don't jump, don't jump." Some tore at their hair as they looked up and saw children dangling from the windows, held by little friends inside the building. Students in the classroom begged these classmates to come back into the room, as the smoke got thicker and blacker. We heard those mothers crying, "Don't jump," but it was getting so hot the window sills were blistering and we heard the crackling of the enameled wood.

Two men ran off the street up the annex staircase (which was the only metal staircase in the school) and reached a window that was kitty corner from the window where some students were perching themselves, trying to land on the roof of the porch. The men could not reach the children. The windows were too distant. A ladder was finally placed on the roof of the porch against the outside of the north wing. Children could barely reach the top of the ladder and balance themselves while the men, one a priest and the other the father of one of the girls in my classroom, frantically pulled the children up the annex window. The annex was filled with smoke and the men had to grab the terrified youngsters and push them down the metal stairwell.

They fell, stumbled and pulled each other down the stairway. They were dazed in the rush and fury, and fell to the first floor landing. But through the haze of the smoke they could see a clearing of daylight and this impelled them to go the last eight difficult steps. They saw crowds filling the streets. They saw students like themselves, but somehow different now, because for the first time they truly saw life. They would never again forget what it was to be alive. They were the living.

While they escaped into the harsh coldness of winter, destruction and death were continuing behind them. The latest survivors began searching for other survivors and looking for familiar faces in the crowds, looking for their brothers, sisters, and cousins.

The panic in my room mounted. My girlfriend Helen rushed to me and ordered me to cover my face with my sling and give her my handkerchief. Tears were pouring down our faces. I gave her my hanky and we both covered our mouths with cloth. I blew softly into the sling around my mouth trying not to breathe in the hot gases which were filling the room. Helen went toward a window.

I struggled to again get near a window but the mass of panicky children trying to climb high and higher over each other's bodies held me back. A big boy grabbed another boy by the throat and forced him away from the window. This was useless, because another struggling student would fill the space. The crowd was constantly shifting either by falling from a location or being pressed into a new position.

Arms, heads, legs had no beginning. Girls and boys were twisted grotesquely as they pushed one way and then another, all begging for air. I was coughing, as most of us were, and everyone was crying. Tears streamed down out cheeks. I wiped my face and it was oily. I grew weak and students crushed me until finally I gave up struggling to be by a window because I found I could no longer breathe there. The smoke was as thick by the windows as it was in the room and I could not stand the pressure of struggling bodies pounding me any more.

It was growing silent, the chant "fire, fire," had stopped, the steady beat of the fire alarm now called attention to the danger. Pleadings from the children now were for "air."

I went back to a window and saw the firemen come. They were

standing behind the iron fence beating on it with a wooden ladder, but they could not tear the fence down.

The firemen had come! The firemen could not reach us! The ladders were too short!

I coughed and I cried as my legs grew hotter. There was a girl named Lisa begging for air so I let her take my place at the window. I was so tired and it was a struggle to change places with her. She pounded on the pile of people forced into the opening and she managed to squeeze her way to the window but soon gave up her place to another hysterical student. Lisa stood by me and screamed and cried, "Help, let me out!"

All of a sudden there was something covering my back. The blackness of the room no longer mattered; not being able to breathe was no longer important to me. I was being attacked by a wild monster who was pulling off my skin. I jumped and grabbed for it and tried to pull it off, but it kept tearing and gnawing away at my back. Suddenly, I was being pushed down by heavy weights. A bright, orange doll appeared and curled up in the most grotesque manner. The straw doll and I were falling and the room was getting smaller and smaller, but the monster on my back was still devouring the skin off my body.

I screamed.

The "straw doll" was Lisa. The monster on my back was fire. I catapulted myself out the window and I felt the pit of my insides wrap around me as I fell crashing to the ground.

CHAPTER III

The first thing I remember was seeing the dark pants legs of men racing me down a hallway. They knew the right direction to take, but I did not know where we were all hurrying. I realized I was in a hospital, and this did not frighten me, but there were strange things happening to me.

I was a child but I did not feel like a child anymore. No one around me understood this. I had traveled to the limits of experience where I had never been before and beyond. I saw children hold on to the pinnacle of time and fall off by a nail's edge. I had witnessed an intensity of peace and beauty, but now the feeling was gone forever. The journey into self was frightening but the experience of coming back and staying was more difficult to accept. I was trapped into a system over which I had no control, and I wanted to escape from it.

This was more than just a desire to escape the agony of the burns; it was the beginning of a listlessness in living that has never really completely left me. The acceptance of dying began a few moments after the fire. This acceptance of death was often to lead me to be restless at living.

Ceiling lights whizzed by me as I was being pulled very fast down long hallways. I felt extremely confined. There was a pressure around me making me feel that my feet were higher than my body. Everything around me was blurred because I had lost my glasses in the fire and my eyes were nearly swollen shut. I heard commands and orders being given and I wanted so much to ask them to stop for a moment, but I couldn't get anyone's attention.

Suddenly we came to a jerking halt and I could feel the rawness of my burns entwined around my whole body as my seared skin pulled away against the sheet in which I was cocooned.

A familiar sounding voice shouted, "Stop, this is a bad one," and I felt the presence of someone standing over me and saying something. Through blurred vision I could make out the purple ribbons over my head and then I understood that the familiar voice was one of the parish priests giving me the last rites of the church.

I wanted to stop and tell him that I was not a bad girl. I was confusing the condition of my soul with the state of my physical condition. I hoped someone would tell him I was not bad, but I couldn't get anyone's attention, because they kept hurrying me down the hospital hallways at a fast pace. I felt pain with every movement of the stretcher, but that was not what was causing me so much distress.

The race continued to the sixth floor operating room and in there I knew I was alone, even though there were many people. No one knew me. This frightened me because I felt that I was getting bigger than the room. Suddenly I was crawling around inside my body, not resting in it, but creeping around looking for someplace to be, but I could not find a safe place.

I begged for something for the pain. I could feel my legs were about to burst, and I could not think why. I never thought at that moment that I was burned, I just needed something for the hurt.

It was cold in the operating room and there was a constant stream of water running someplace. Over my head was a big, bright spotlight and I thought it was falling on me. I was afraid it was going to crash down upon me and I would start falling again and never stop. "Please don't make me fall," but I could not tell these strangers what was happening to me. They did not understand where I was.

Everyone was dressed in green scrub clothes and busy doing things around and to me. I was picked up and placed upon an operating table and it hurt me to be moved. I was wincing in pain, and wanted someone to talk to me, but they were all grabbing parts of my body and no one paid attention to me.

I kept telling them to get my mother. I wanted my mother. I gave them the name and address and my girlfriend's telephone number. My mother was deaf, I told them, and could not use the phone. I knew she would be worried because I didn't come home from school.

But school was not school anymore. It was fire and children screaming and burning around me.

I could hear the quick, sharp, snipping of the scissors cutting off my uniform skirt. This embarrassed me. I really did not want my clothes taken off before all those strangers in the room. I was having my period and everyone would know. I felt ashamed.

"Please stop the pain," I begged them, but by then I did not know what pain I was concerned with.

I was outnumbered and I could not tell anyone what happened. I was lonely. I knew there were lots of people around, but I could not see their faces since they were covered with surgical masks. This frightened me. I could not tolerate being handled by all those strangers.

I kept on drifting right inside my body like I was crawling around in it, and not able to leave it, but still not comfortable in it either. I felt that I was watching what was happening in the room, but the anxiety of being alone with all those people dressed in green robes who were grabbing and poking at me, made me cry. Then someone stuck a needle in my arm.

I wanted to know where Lisa was, but no one knew her, or me. I was not in the body they were working on. The steady stream of water in the background and the clicking of surgical instruments kept ringing in my ears. I could hear authoritative voices calling out orders. I could not tell anyone what happened; they would not understand my flight into hell. I was cold and they were stripping me, but I wanted to be covered and someone to hold me, to stop me from falling. My feet were getting higher and higher, but I was falling again. I was crashing again.

Oh, God, this was hell without fire. No one had faces. No one knew me. I was crying. I was bad. I was a serious one. Streams of liquid kept pouring down my face. I was bleeding. I wanted my mother.

I was alone.

I was dying.

Part Two

The Parents' Search

CHAPTER IV

About ninety minutes before the fire began, the nun in room 211 selected twelve boys to go to work on the clothing drive in the basement of Our Lady of the Angels Church. The boys walked down the hallways that minutes later would be an inferno.

Around two thirty, two boys were in the basement of the north wing of Our Lady of the Angels School's furnace room located in the chapel. They were performing the daily routine of emptying wastepaper baskets for their classrooms. The boys smelled smoke but could not determine its location. They immediately returned to their second floor annex classroom and reported their suspicions to their teacher. The teacher conferred with a neighboring teacher as to evacuating their classrooms.

One teacher ran to the principal's office in the south wing of the school, past the fire alarm located on the side of the archway door. The fire alarm looked like an ordinary light switch, the only difference was that it was placed much higher, so that no one would ever toy with this important alarm. The alarm was not connected to fire department headquarters, it was only a warning alarm for those in the building. The teacher discovered the principal's office was vacant. The principal was substituting for a first grade teacher who was ill, and had left her office unattended. The teacher ran back to her homeroom and both rooms 206 and 207 were evacuated.

THE PARENTS' SEARCH

Outside the school a passing motorist saw smoke coming from the school; he ran from his car into a store shouting that the school was on fire, but the storekeeper was frightened and confused and answered, "There's no public phone." The driver left the store to see shooting jets of flames coming from the rear entrance. He ran down the street knocking at doors of houses, but the first homes he went to did not have a telephone. In the meantime, the storekeeper had run out of the store, saw the fire and smoke, and used the store's private telephone to call the fire department. By that time the school janitor, who was working at the kindergarten building down the block, had seen smoke and flames. He ran toward the building and shouted for the rectory housekeeper to call for help. The frightened housekeeper called the fire department and reported that Our Lady of the Angels Church was on fire.

Inside the school, two girls who had been acting as teacher's aides for the absent first grade teacher returned to their homeroom 211, the eighth grade classroom right next to mine, to report that they had smelled smoke on the first floor. Instead of leading her class out of the building, the nun ordered her students to kneel down and pray the Rosary When a girl went to the rear door of the classroom and opened it, the force of hot smoke in the hallway pulled it out of her hands. Smoke poured into the room. The kneeling students abandoned the prayer and the struggle for air began in room 211. All the windows in this room were twenty-five feet from the ground, just behind the courtyard's iron fence. The students in the room had no choice but to jump, landing on the pavement twenty-five feet below, or wait for the firemen to come. Most of them had to wait too long, including one of the girls who had reported the smoke. Out of the forty-five students present in the classroom at the time of the fire, more than half were killed.

Minutes passed, and the school fire alarm still had not sounded. While everyone at Our Lady of the Angels knew how to march in a fire drill, few knew how to report a fire.

THE PARENTS' SEARCH

Minutes escaped. The school fire alarm remained silent.

This failure allowed the fire to triumph and gain the momentum it needed to spread as precious seconds ticked away. The fire traveled faster up the back open stairwell of the north wing. In its silent imposing manner it raged up the stairs blocking the hallway exits. The extent of the deadly force of this inferno was not immediately known. Its murderous gases were not noticed until the fire was so out of control that it hungrily, furiously claimed lives in its wild path.

The flames licked over the varnished wooden staircase and sucked up the plaster wallboard, causing more deadly gases to spread. They went unnoticed until students in several classrooms felt the heat outside the doorway. They noticed the faint smell of something bizarre, something strangely amiss in a school. No one believed it could be the smell of smoke. It was winter time. Not the time of year when sometimes the pungent, pleasant smell of burning leaves wafts in through windows open to the autumn air. No, it is cold outside and the windows are shut, we hear only the hissing of the radiators. But we started sniffing the air like little wild coyotes trying to confirm that there really was something burning inside the school.

Survivors from the rooms which faced the alley told me the panic of the fire was heightened by the intensity of shouting "Fire, fire," out the windows to a quite vacant alley. You could see only the roof tops of the neighboring houses, but there was no one on the street to hear their desperate calls of "Fire, fire."

A girl told me that as the isolation built, terrified students ran toward the windows as they saw the classroom filling with toxic gases and growing darker and darker. Small explosions were set off. The globes of the electric lights burst apart. A boy sitting behind her fell over, his head thumped hard on the desk and she screamed, "I got to get out of here, I don't want to die, my mother needs me."

Then, at last, the familiar sound of the fire alarm burst into action and the steady bong, bong of the warning could be heard through the rising smoke. Still, the firemen were not there.

Finally there were people down below the windows screaming, "Jump, we'll catch you." Parents lay prone upon the pavement and begged the students to jump on them. Many of the children tried to follow these instructions but missed the protective pile of men and women offering their bodies to break their fall, and were badly injured. These children were immediately snatched up, pulled away from the school building, and dragged out to the curb. Some were dead, others were still breathing but their bodies were covered with charred, blackened, swollen skin from the intense heat. Some others picked up these little cinder-bodies destroyed by the fire, held them close and tried to comfort whatever child was lying limp in their laps.

A student jumped. The child landed on the pavement, but could not move his body. His head rested on the hot building and the cold winter air felt comforting. Someone threw a blanket over him; the child screamed because it hurt to be touched with the weight of the blanket. Two men carried him away. He looked up and saw the thick smoke pouring out of the window where he had stood minutes before.

Other mothers kept on running up and down outside the relentless iron fence and yelled through throats hoarse from continuous screaming, "Don't jump." But it was getting too hot for children to heed such warning. Wooden window frames got hotter and hotter. Radiators were getting too hot to touch. Skin on little bodies was getting too hot to touch.

Students were pushing and jumping out of windows before help arrived.

In a fifth grade classroom the teacher tried to encourage the little children to jump. Some of the students sat frozen in their desks afraid to move. Some classmates told each other it was safer to jump. One little girl pushed her friend to safety but fell back into the smoke filled room.

Frightened students jumped or pushed each other to the cement pavement below. Finally the sirens of the fire engines could be heard!

The fire department arrived just a few minutes before school normally would have been let out for the day.

Engine 85 clanged toward the direction of the church, as the first alarm given to the fire department was Our Lady of the Angels Church was on fire. The maddening crowd ran after the fire trucks shouting, "The school's on fire, over here, over here." The fire engine backed down the street through the crowd.

One eighth grade boy who was working on the clothing drive walked out of the church basement and saw the commotion and ran back to tell the pastor that the school was on fire. The priest raced madly toward the school and a fireman yelled at him, "Don't worry, Father. All the children must be safe out on the streets. The children who could walk out of the school are being directed to go into the church."

Engine 85 arrived at the school. The first response to a fire is a Still Alarm; this includes an engine to pump water, a hook and ladder truck, a squad and a battalion chief. The fire was out of control, there was not enough man power to help the trapped students. Immediately a Box Alarm was called. The firemen quickly set up wooden ladders on the alley side of the school building. The ladders were too short. These wooden ladders were twenty to twenty-four feet long and were used in emergencies on maybe first, or low second floors, also for the bridging of fences and so forth. The longer aluminum ladders were on the truck and there was not enough man power to raise them up to where the students were.

Firemen raced to reach children. By the time the second engine company arrived the iron fence was still standing, holding back the rescue workers. There was a girl hanging outside the building by the edge of the brick mortar. Her classmates took turns holding on to her. People shouted at her,"Go back into the room; you might fall." She managed to hang there until a fireman reached her.

The fire went from a Box Alarm, to a 2-11 calling for more

equipment. The firemen were frantically beating the iron fence trying to strike it down, but the fence wouldn't move. A second company arrived and a hook was used to tear down the iron fortress. The firefighters rushed into the courtyard and threw up ladders to the trapped eighth grade students. The wooden ladders were too short and too few. Students screaming for help called out to a fireman hoisting a ladder, "Over here, over here;" the firemen pulled the ladder from one window to the next window.

Nets were dragged into the courtyard, but more manpower was needed to efficiently use the life-saving device. Nets were quickly abandoned. Parents were begging children to jump into their arms. Some were caught.

A 5-11 was called, the maximum all unit alert. A section of the roof collapsed and a new piece of equipment arrived, a snorkel. This was a giraffe-like support able to raise firefighters with hose lines closer to the roof of the burning building to douse the fire.

Firemen worked at top speed snatching children from windows and dropping them to be caught in the arms of other men. But the fire was too hot; a window full of children waiting to be lifted out was engulfed in a burst of flames and all those children, begging for help, were incinerated.

How totally useless those men must have felt, to be caught in the miscalculations of disaster, to be trained as fire fighters and have to stand helplessly by and see children trapped in windows by flames and smoke and not be able to reach them. How terrible to be fire fighters and not fire protectors.

CHAPTER V

Television, radio, and newspaper reporters rushed to the scene of the fire. They stood among the crowd trying to interpret to their audiences the burning of Our Lady of the Angels School. Radio and television broke into their regular programming as news flashed across the air waves.

Students who marched out of the school by the prescribed method of fire drills were herded into the church and told to wait there. However, they were soon told to leave and go home when the battalion chief warned that the church might also go up in flames. Panic was spreading. Nuns and teachers warned the children to go home. They ordered their classes to go to neighboring homes and ask for blankets to wrap themselves in and to go straight home.

Out on the street little children, frightened by the excitement and noise, cried for their mothers. Strangers in the crowd picked them up and held them close and told them everything is "okay." They wrapped the children in their big coats and took them into their homes all the while holding their hands over the eyes of the little children to keep them from seeing the panic around them. The children were too young to see, but it was the children who were suffering.

Families stood on the street corner looking for their children.

THE PARENTS' SEARCH

Students covered with smoke and grease, who escaped from the eighth grade classroom, gathered on the street and watched the terror they had just left. They stood in a daze as though watching a scene from some horror movie, while people walked up to them and grabbed them and started kissing them shouting, "Thank God you're alive."

One girl, covered with blood from cutting herself on the broken window pane, was thought to be alive. "Yes, I am alive," she whispered, but then she saw them dragging victims from the courtyard and screamed "Get away, get away from the dead bodies." She was carried into a car and driven to a hospital.

The eighth grade boys who were working in the church basement at the time the fire broke out stood on the street corner with the parents of missing students and said they would look for their sons and daughters. They tried to comfort the parents among the throng of people but would turn away when they saw the body of a classmate being carried away from the school. They went into shock when they recognized the lifeless body of a friend who minutes before they had been sitting with in a classroom.

Everybody on the street was wild with passion, searching for children in all the confusion. The command was to go home.

Steadily the stream of bodies began piling up in front of the rectory. Police tried to keep people back but could not fight the crowd. It wasn't a normal angry crowd of people, but a dazed hurting group of people who were in panic at the fear of losing their children.

Large crowds quickly gathered at the school. People from all over the city came to see the torrid monster and watch the neighborhood look for its own.

Parents rushed to children and started shaking them, pleading, begging the confused students to tell them the whereabouts of their sons or daughters.

The weakened burnt children were lying on sidewalks covered with blackened swollen skin ready to burst. They were quiet and weakly asked to be covered from the December cold. The most critical ones were wrapped in sterile sheets and rushed to hospitals either in ambulances or paddy wagons. Private cars were rushing the less seriously injured to the hospitals.

Priests walked outside the school annointing the injured and the dead. Many of the children had been baptized by the same men who were giving them the final rites of their church.

Fire hoses formed large mazes of rubber, spider-like webs. Water soaked the people as they milled through the streets looking and watching and climbing over the big thick hoses. Mothers ran wildly up and down the street and then suddenly grabbed onto their missing child, screaming, "Thank God, thank God." They held onto their babies and in all the commotion around them stopped and rocked them gently in their arms. The seventh and eighth grade boys who towered over their mothers were embarrassed and said, "Stop it, the fellas might see you." But it didn't matter because life was another color now.

Outside the priest's house the lines of canvas bags which contained dead students were getting longer and longer. Police stood guard over the bodies, pushing back mothers and telling them to go get their husbands or look for their children in the hospitals. Nuns knelt on the street with the dead, praying over the children they had once taught. The nuns who had taught the students their prayers, were challenged so many times that day as they prayed to God for the dead and the living.

The ghastly, pathetic lines of dead bodies grew longer as firemen continued in carrying the victims toward the priest's house. Ambulance drivers backed up the vehicles and placed the bodies inside, helped by the parish men who held back weeping mothers looking for their children.

For many of my neighbors, the search went on all during that long night. Mothers holding infants in their arms pushed through the crowd of curiosity seekers, on choked off streets, and cried out, "Let me through. My child is in that fire." These women were joined by grandfathers who tried to hold them in their arms and comfort them by saying that probably the children had gone home by that time and were safe. Fathers dressed in work overalls held on to each other crying and hoping that somehow their children had managed to escape, and then they began searching through rubble and the confusion of fire equipment for those youngsters.

A neighbor, whose own children were safely at home, went to the aid of a friend who was standing and screaming, "I want my baby," and said to this tormented woman, "Maybe your child is in the hospital. Go check the hospitals," or "Maybe your child got confused, and is lost in the crowd, or maybe going home at this very minute."

But, in many houses that evening children never appeared. Women in housedresses and sweaters wiped their eyes on their kitchen aprons and kept on searching. Street lights were turned on. The Salvation Army set up relief headquarters and a loud speaker system. A voice began booming over the din of the crowd advising people to "Go home. Go home." Later the Salvation Army volunteers gave out directions on how to get to various hospitals and offered rides to anyone in need. They told people to go to the rectory where hastily prepared lists of fire vicims were being posted. Parents walked to the rectory and many of them were so terrified at what they might learn that they could not go up the three steps without help.

Children, with fire smirched and torn blue school uniforms, mixed in with the crowd and tried to help locate sisters and brothers or cousins. Sawhorses, set up to hold back the crowds, were pushed closer and closer to the actual burning building.

Parents, trembling with fatigue, persisted in the search and became lost in the crowd, but still hoped that the missing child would miraculously appear and make them a family again. They ignored the admonition, "Go home."

To go home was useless, because the home was empty, the children were missing. The dread of going home and finding the truth was too much to bear. These men and women hopelessly kept stopping strangers and asking, "Did you see my child?" But, they received no response except the monotonous litany of "Go check the hospitals." The hospitals became the last hope of the night. When that search came to a fruitless end, there remained the final act, the long journey to the morgue. There many of those fathers and mothers faced the most bitter loss of their lives, the unexpected death of their healthy, lively child—now gone forever.

CHAPTER VI

A police captain went across the street into the convent to use the phone and call the coroner's office requesting that more doctors be rushed to the scene. "There's too many victims to send to the hospital just to be pronounced D.O.A. (Dead on arrival)," he exclaimed. "Get someone down here so we can separate the victims and rush the injured to the hospitals and the dead to the morgue."

As he walked out the convent door, the policemen looked upward and felt sickened as the black clouds billowed and engulfed the school building and filled the skies all over the immediate neighborhood with poisonous gases. As the deadly fumes filled his nostrils he coughed and gagged and wanted to throw up on the street. The thick heaviness of the air, and the mingling stench of burning wood and plaster, polluted the crisp winter air.

By three thirty the fire was smouldering and over, the damage was done. The saddened Fire Commissioner delayed calling in the "Fire struck out" (a term meaning the fire was out) until four seventeen, hoping that the delay might enable his men to find more children alive and well—but it didn't.

He ordered his men to begin the heartrending task of entering the building with body bags and to start searching through the smouldering rubble for the remains of the trapped students. The bags would hide the pathetic contents from the eyes of grieving parents and curiosity seekers.

The firemen stumbled through halls and up stairs which just a short time ago were normal exits for school children, but now were littered with a collection of wet falling plaster, charred wood, smashed walls, and dead bodies.

Searching through the still smouldering, charred debris they found some of the children slumped at their desks almost as though gently sleeping, and others who were no longer recognizable as children. The stalwart men wept at the pitiful sight.

Daylight was fading and floodlights were set up to shine upon the north wing of the school standing gaunt, blackened, and useless. Firemen continued running into the structure and returning with ladened coroner's bags which were gently laid in the rapidly lengthening rows on the street near the rectory. Priests walked among the pathetic bundles, annointing the dead.

The fire was out, but the Fire Commissioner still delayed giving the "Struck Out" signal. Firemen toiled on locating more bodies in the classrooms of that north wing, digging through mounds of fallen plaster from caved-in ceilings and under desks. They came upon little bodies holding onto each other. The men kept on working and carrying each youth tenderly wrapped in a bag out of the ruined building. The men fought back the tears scalding their already smoke reddened eyes, but still they kept on with the search.

The building was dark and the men needed searchlights as they pulled open doors to schoolrooms which an hour or two before were full of happy, noisy, chattering children and nuns. School books were open, some not even charred by the fire, and pencils and paper were scattered on overturned desks.

One fireman found a golden-haired little girl curled up on the floor beside her desk as though quietly sleeping. The burly fellow drew out the big bag, picked up the small child and held her tenderly in his arms for a moment as though trying not to disturb her further . . . and prayed. It was not the prayer a devout Catholic usually recited for the dead, instead he found himself fighting with his God, yelling at God for permitting such a horrible thing to happen.

He wept as he knelt and tied the straps of the coroner's bag. He tenderly picked up the small body and effortlessly carried his burden out of the room. His knees grew weaker and weaker as he started out the doorway and walked towards the stairs, that by some miracle were intact and not fire damaged. As he drew nearer he looked up at a statue which was ashen with smoke, but by some chance had not been burned by the flames.

He recognized the statue, famous as the "Sacred Heart," a replica of Christ standing with arms outstretched and reaching to the sky. Upon the chest was a huge heart with plaster rays extending from it and the fireman thought, "Oh, Lord, why did you need them? They are just children."

He walked on down the stairs and at the door encountered confusion. Firemen, policemen, and friends struggled to hold back a man who was beating against them and shouting that his daughter was trapped inside and he was going in to get her. Two firemen bodily lifted him and carried him to the curb, restraining him, telling him he could not go into the school. "Go on home, go on home," they all kept advising the man, and sat with him on the curb until he stopped struggling.

A volunteer from the Salvation Army came over and embraced the man and asked if he could help him to his home. The volunteer, dressed in his dark blue uniform piped in red, sat with the grieving father and patted his shoulder. Suddenly the man stood up and said he was ready to go home and look for his daughter.

The volunteer went with him, walking not alongside but a little behind, just staying there giving moral support to a frantic father of a missing child.

At last, the fire was called officially "Struck out." Darkness of the midwest winter night settled down. The street lights were turned on and illuminated the benumbed faces in the crowd of people too paralyzed to move away from the terrible scene. "Go home, go on home," the voice still pleaded over the loud speakers, but the gaunt-eyed women, holding infants clutched closely to their hearts, could not move towards their houses.

In city wide factories those husbands and fathers punched time-clocks, gathered up worn metal lunch buckets, and prepared to go home, unaware that they were to face homes broken apart by tragedy. These unsuspecting men unlocked their car doors, turned on the ignition key, and flipped on the car radio to catch up on the evening news, only to be startled to hear a voice different from the usual newscaster. This announcer was giving a report on a fire and instructions about areas blocked off. He was saying, "South on Madison, east of Central, west of Kedzie, and north on North Avenue, all routes were closed to traffic. The area is being kept clear for fire apparatus, and other emergency equipment, and ambulances. Notice! All cars are to stay away from the vicinity of Our Lady of the Angels School. There is a four alarm fire. Children are trapped on the second floor. Please stay away from the scene of the fire—The extent of the damage is not known, but fatalities are mounting. Repeat. Stay away from the areas south on Madison, east on Central, west on Kedzie, and north on North Avenue. Please keep the area clear!"

Fathers with children attending Our Lady of the Angels School drove hysterically as close as possible, then left their cars standing and ran towards the still smoking building. They clawed their way through the growing crowd of people and joined the women standing in the cold staring blankly at the ghostly building.

Children were interviewed by reporters and asked how they had escaped from the school. Few were actually in the north wing when the fire happened. The children were excited giving accounts of their escape. Newsmen hurried to pay phones located in empty stores in the neighborhood. They ran into drugstores and grocery stores and no one was there; everybody was gathered at the school building trying to locate their children.

Photographers rushed to the scene and took pictures of the grieving parents and the rows of bodies. They caught the hysteria of the afternoon in glossy black and white prints. These photos would be engraved in the memories of the cameramen for the rest of their careers.

All the professionals involved with the details of this disaster were stunned: the rescue workers, the reporters, the police. Many were parents themselves, each wondered how could such a dreadful thing happen to innocent nuns and children.

The mayor and the coroner stood in the crowd, watching the faces of the people, and wept. Everyone dazedly wondered how such a dreadful thing could have happened. Many rushed to help parents who had begun the agony of searching for their missing children.

CHAPTER VII

People started arriving at the morgue after five o'clock. They had already checked the hospitals and their children were not there. Parents asked did they have their son, their daughter? She is nine years old wearing a blue uniform. "No, there are no identifying marks." They met other parents waiting in the hallways of the morgue. Shouting and shoving for attention, everyone desperately wanted to get hands on the lists of little victims who were there. Attendants at the morgue tried to be heard over the noise because they needed more information about the children before they could compile a list of names to announce.

Identification was going to be difficult, they said, because some bodies were burned beyond recognition. It would take special identifying marks to help name who was arriving at the morgue. Family dentists were contacted to come to make identifications. "If there are any fathers or male relatives of the children present," they said, "please stand in line and you will be permitted to see the children with an attendant." There were doctors and nurses at the scene to help calm parents who found and recognized a child and cried out in agony.

The tensions mounted and people were not willing to line up in order, they were mad with uncertainty. Ashen-faced, gaunt-eyed mothers waited as fathers, uncles, and grandfathers went down stairs to see if their little ones were among the multitude of children draped in canvas. They screamed as covers were pulled back, but the bodies were burned beyond discerning whether they were even a girl or a boy. "Perhaps some jewelry the child had on, a ring, a bracelet. Did the child have a holy medal that would aid identification? Would you bring down the mother, maybe she could recognize a piece of clothing, a sock, a shoe?"

Fathers would come up and look for their wives in the crowd and hold them close, and say, "I can't find Johnny, I can't find Johnny. I saw Billy, the boy across the alley, he's down there, but I can't find our Johnny."

Parents who could make positive identifications were called over to a desk to fill out a history of the child. A man stood there and said, "Sorry, but I got to do this for the records, could you please answer a few questions?"

This went on countless times, all through the night. "Name, address, age, married or single, race, birthplace, birthdate. How long in the United States, how long in Chicago, Mother's maiden name, present occupation, employed by whom, past occupation, social security number, cause of death, cause of occurence, place of occurence, place of death, Cook County Hospital DOA. Date of Inquest, December 10, 1958, Date of Death, December 1, 1958 Hour 2:45 P.M. Mental condition.........., physical condition.........., sight.........., hearing.........., crippled.........., deformed.........., height.........., weight.........."

The staff kept apologizing for insisting on information that was really meant for the old, not the young, but it was for the record, they said. People were then instructed, "Go home and contact your family undertaker. The arrangements will be made to move the bodies to funeral homes."

Mothers and fathers sat, some were chain smoking, others were sitting, twisting and turning workcaps held in their hands, trying to keep their minds off the horror of the night. Some people tried to comfort each other, while others sat aloof from the crowd not wanting to be disturbed. People greeted each other with shocked looks on their faces. Some stood in hallways fumbling in their pockets for loose change to drop into pay phones to call home making sure the children were still missing and had not suddenly appeared. They had gone from hospital to hospital, and checked all their neighbor's homes, and then finally summoned the courage to drive off to the morgue.

Sometimes, someone said "God be with you, they are little angels now."

A mother stared into space holding onto her purse and a child's jacket she'd carried all night hoping that the child would be found in a hospital and could be taken home. Now she carried the jacket because she had nothing left, just the jacket of the child.

At last there was nothing to do but to go home, park the car in the garage, and climb up the back steps. Lights were on in all of the neighborhood houses. People were waiting, waiting for news. Neighbors came to the homes of the bereaved families and made countless pots of coffee. Grandmothers stood in the kitchens inviting friends and relatives to sit down and have something to eat. They needed to keep busy to push out of their minds the reason families were gathering.

Mothers held living children close to them and thought of the captive child in the north wing. It was worse than anything they had ever endured.

"A child does not go to school to burn to death. Is there not a heavenly Father who protects us? He could not will such a horrible thing to happen. God is just, God is good, but why is my baby gone, gone today, forever? Did I tell the child I loved him? Did she know that she was loved and wanted? I am a bad parent and this horrible deed was my fault, because I did not do enough to show my love. I shall never let the child die in memory. I will carry this child with me all the days of my life. I'll make sure to tell my child I loved him, but why was the child taken from me? I am the one who should die—not my baby."

Part Three

Suffering Has No Teacher

CHAPTER VIII

The walls were closing in around me and I knew I was in bed. It was dark in the room and very quiet—at last it was quiet. I could not move any part of my body. I wanted to lift the heavy weights off my chest, but those weights pressing down upon me were actually my hands bandaged like two white clubs. These clubs were pinning me upon the bed. I was exhausted, but I wanted to stay awake. I wanted to know what happened.

I remembered being in school and the school burning down, but the fire seemed so remote I pushed the idea out of my head. I kept thinking about all that had happened and I was still not certain about being in my own body. I wanted to remember, I wanted to forget, but no matter how my mind raced around, I could not deny what had happened.

I was certain there had been people around me dying, and now I was alone in a bed somewhere; I was not sure where or even care. I was too hurt to be concerned, but I was still conscious of what was happening around me. I sensed I was alone in a small cubicle of a room in a hospital ward, but this time I did not mind being alone. It was so much worse being alone when I was on fire and there was no one to help me. It was so much worse being alone while hurtling out of a window. This kind of alone was better. I was not falling; I was not burning. I was tightly wrapped in bandages. I just wanted to stay and be quiet where there was no more pushing and no more struggling. Suddenly the solitude of my room was broken by the sound of uncontrolled sobbing. A figure was bending over my bed crying and praying, "Oh, my God, please, God!" I could recognize the voice. It was a man, a Jewish man. I had never heard a man cry before or a Jewish person pray before, but now it seemed completely natural. This was a former neighbor. His daughter and I had grown up together until she moved away in the fourth grade.

The man was crying hard, so I said, "Don't cry. I'm all right." Little did I realize the terror of that day, how families and friends would suffer over the loss of lives and the anguish of seeing children suffer.

I felt my face dripping and liquid running down my neck and back, but I wasn't crying. I couldn't turn my head, however I knew who was with me. He cried and prayed and prayed. It was nice to hear someone familiar, although he was from the past. I wanted my family.

The next person at my bedside was a lady who screamed, "No, this isn't Michele," and then ran out of my small room weeping and calling that I wasn't Michele.

"I am Michele, I am Michele McBride," but I couldn't call out, and I couldn't stop the strange woman from disappearing. I did not know her, but I wanted her to know I was Michele. I was weak. Again I was losing control, and I could not stop what was happening around me. This time being alone was scary because people did not know who I was.

A long time later I discovered this woman was looking for her daughter, another Michele. However her little girl was never found in a hospital.

Then my mother and father came. I recognized them by shape and figure before they even spoke to me. Some other people were with them, I didn't know who, and it didn't matter. All that mattered was my family had found me. They would tell me what had happened and everything would be all right. I would not be alone any more.

I wasn't the same little girl who had left for school that morning. I was experiencing life at a different level now. Suddenly my family were strangers to my past; my future was uncertain. My experience was unique, and nothing could have prepared them to understand all that had happened to me. I was separated from my body, and now my family. I had to rally. We were all desperate, trying to protect each other from more hurt. The only good thing about being so seriously injured and finding me in the hospital was that there was hope. I was alive. This was little sanction that my parents would not have to make the grim pilgrimage to the morgue.

My father was the first to come near me, and he was grief-stricken to see me burnt beyond recognition. Only my voice gave hint of who I was.The wounds of a burn victim can overwhelm the most experienced medical team, but for the unsuspecting family the toll is greater. There I was with my face blackened, my head swollen three times its normal size, hundreds of stitches running across my forehead and skull, my eyes almost burnt shut, my eyebrows and eyelashes gone. I had tubes through my nose to a stomach pump, a catheter in my bladder, blood and I.V. feedings flowing through tubes in my feet. I was able to recognize familiar color, and I saw my mother carrying my beige Sunday coat and a pink, baby blue and egg-shell white multi-colored blanket she had made for my baby brother.

Mother had thought she could come and take me home that night. But as she walked into the room and saw me she fell against my father. They both were crying, and the group of attendants at the foot of my bed cautioned them in whispers not to cry as this might frighten me; I was awake and aware of what was going on around me.

She came close and leaned over the bed and I was happy to have her near me; I would now be able to tell someone what had happened at school that afternoon. All the confusion and the horror could be shared with my mother who was with me. I would no longer be lost in a crowd who did not know who I was. I could not possibly be avoided and ignored now. Mother would pay attention to me.

As Mother stood close to my bed I asked if she could read my lips. She nodded and I formed my words carefully so she would have no trouble understanding what my lips were saying. There was a tightness of the skin on my face, my lips were dry, and I felt them crack and sting as blood seeped out when I spoke.

But neither this discomfort nor anything else could keep me from speaking. I do not think anything could have kept me from talking, not ever. I started to pour out what happened. I told my mother what had bothered me most was no one assumed authority, no one tried to cope with the situation. I was angry, and I could not comprehend why all the things that went wrong had been permitted to go wrong.

I told my mother how illusive the situation had been, how grotesque it was, sitting in a school that was being consumed by a raging inferno and hearing the command to recite the Rosary. "The dumb nun said 'Pray the Rosary,' the longest prayer in the history of the church," I fumed to my mother. I could not get over the ridiculous command. It shocked me. Perhaps this was all to the good, for it shocked me into anger. Anger in turn had moved me to action during the fire and saved my life.

Up until the fire, my experience through books and movies had shown that when disaster strikes there is order, there is bravery; only the villians have to struggle. Perhaps I was brainwashed by the movie "The Titanic," when people stood on the decks singing fond farewells. I think I would have responded better to a cheer song than to the Rosary. That command had been shocking, and I had moved away from the escape window to get away from the madness.

As I lay in my hospital bed, I knew full well no one who ever walked into my room could actually comprehend the horror in the north wing of Our Lady of the Angels School.

The long charade to keep information from me began. The

aloofness of dying kept me separated from those around me, because I was suffering a dimension of life not shared by everyone, especially at thirteen years of age.

Of course it was for my own good, they said, but my struggle to find out what had happened began.

Neither my family nor the medical staff would confirm for me what I was certain of. I knew Lisa was dead, and I knew others had died because I saw them, however no one would tell me so.

I wanted to feel a comradeship with other children who survived the fire, but those who had been around me as the room burned had either died or gone into shock because of the atrocities we witnessed. I needed desperately to find out what had happened, to confirm for my own peace of mind and sense of being, that I actually went to school one day and the room caught fire. A tiny little room in a corner of the universe had gone up in flames.

I needed to know who had lived and who had died, but no one would tell me.

I do not know whether my horror of the fire came from its immediate power and danger or from its ghastly incongruity. I had been in safe and familiar surroundings with my friends, and the world had gone out from under me. Are fires easier to accept when they happen in more "natural" circumstances, as in a kitchen fire or a car crash? I wondered. I think my overwhelming need for knowledge gave me energy to recover. The longing to understand life and death began and perhaps that is really when I started to write this book.

CHAPTER IX

The hallways were filled with concerned parents, and friends who were bewildered by the awesome bedside vigils of dying children. They lined the hallways trying to find out details and the chances of recovery for their children. Doctors were too busy to stop and console the parents; their first concern was the children.

Parents prayed. Some were shocked when they prayed for a quick death for their suffering children. They felt guilty, but they could not help themselves when they viewed their ruined babies. Babies whose faces resembled hideous rubber Halloween masks, whose oozing bodies stunk, who just a few hours before had gone to school, and now were victims of fire.

Relatives of the children banged their heads and hands on walls because they were useless at stopping the pain. Everybody talked and cried in a special way. Families comforted a stranger to forget their own sick child. They talked with priests, who tried to offer comforting words.

Others sat in stony silence not letting anything invade their personal miseries. They fell asleep sitting up in chairs, afraid to leave the floor for one minute because the condition of a child changed so suddenly. When they heard screams of children they listened carefully to discover if it was their child screaming, and they felt comforted if it was not.

SUFFERING HAS NO TEACHER

People reassured each other about the good medical attention given to the children but stopped and cried in the middle of a conversation, reminding themselves this was really happening. Listening to stories about children who lived across the street and were going to be buried did not make their plight easier.

Everyone envied the parents of children who were not severely injured. Those children were able to sit up in bed, dressed in ridiculous hospital bed jackets, and their parents could kiss them good night knowing soon they would be home.

Going home and receiving wishes from families who had made funeral arrangements was ghastly. People could not be left alone for their private suffering. People pulled through the disaster on reserved strength and on energies that are not mystical but common in distress. Every home had to interpret the disaster, had to form a method of coping. The anger, the love all had to be dealt with. People would busy themselves with the work of surviving, each in their personal manner. The maddening clamour of people suffering was to be heard throughout that night, the week, and sadly enough throughout the years.

I know people who sent other relatives into their child's room first because the parents could not bear to look at the raw, swollen bodies. Looking back on it, I think the parents of injured children suffered as many agonies as the children themselves.

I learned the challenge of suffering is being able to endure various aspects of losses, either physical or emotional—or both. But man can determine his destiny and rise to the occasion. A person alone can brace himself to tolerate great hardships and great obstacles. Even when the torments are as physical and as overwhelming as in this case, a person can find inner encouragement. Every moment he is meeting a challenge, he is growing.

But the harshest aspect of suffering is watching—watching a loved one suffer.

To stand by, to feel, to love, and not be able to alleviate the pain and not to be able to take away the circumstance, is as difficult as the acute suffering. The courage of parents to endure seeing their children turn into hideous looking human beings, forced to hear their screams, their begging, their pleading for mercy from the ravishes of pain is as challenging as the disaster itself. Man has the innate ability to endure himself, but man has to be conditioned to watch a loved one suffer.

Courage spreads if it's nourished, and with the ranks of parents and doctors giving support, most of the children would pull through.

The amount of human energy expended during those critical days was endless. One serious burn is a medical emergency for a hospital; more than one is a medical disaster. Seventy-five children, eleven of them in critical condition, had been taken to seven different hospitals after the fire. I was taken to Saint Anne's Hospital where they had an established disaster program to treat the injured. The most seriously injured were taken directly to operating rooms; the rest went to the hospital auditoriums where emergency procedures could be carried out on a large scale. Hospital staff doctors were called from offices and day duty nurses would do double shifts with the night nursing staff.

Hospitals all over Chicago sent nurses, drugs and blood. Doctors made their rounds in teams all day and all night; some did not leave the hospital for days at a time working under the supervision of burn specialists from Brooke Army Hospital.

The treatments varied enormously, because at that time burn therapy was not an advanced medical field; there were no burn centers. I sometimes feel I was injured during the "Stone Age" of burn therapy. Recovery is unpredictable because so little is understood about how the body replaces lost skin. Most doctors' experience with burns is slight. Many times I heard doctors say their only experience with burns came during their hospital internship as a form of discipline when they had infringed some petty rule!

The effort invested in treating a burn is immeasurable. Many patients, treated for five, seven, nine months, become personalities to the medical staff rather than just room numbers and hospital charts, and then die unpredictably. A successful burn team must treat every burn victim as the first one ignoring the odds against recovery.

Burns are a child's "disease" and this makes the caring so vital. To labor on children whose lives depend on such painful treatment makes it more difficult. Burns are the most avoidable, needless scourge of mankind.

CHAPTER X

I was in critical condition, in shock, losing body fluids, had renal failure, dangerously sick, but I was at no time passively quiet. I was never reverently prayerful. I was angry, and most of all, I was hungry. If dying is a peaceful time, I am sure that is why I did not die in that hospital. I was too loud in my anger and in my demands for something to eat. If I was expected to have the courage to find peace and say a few final words of wisdom, mine would have been "corned beef on rye." I was awfully sick, but that did not bother me as much as when I began to suspect there was a conspiracy going on to starve me. I was so hungry I complained about hunger more than I did about the constant pain from the burns.

If there is something in the human condition that pulls one through impossibly hard times, I am sure mine was being so hungry I didn't realize how sick I was. I was always a very thin child and weighed only ninety-eight pounds at the time of the injury, but I had an appetite that was unmatched.

I kept demanding cream soda pop. I remembered we had some in the refrigerator at home and I begged the family to bring me some. Since I was still in shock, they were not permitted to do this. I demanded to be taken away from the hospital because I thought they really were set on starving me. I kept insisting I be transferred to another hospital—Loretto, where my baby brother had been born. My mother always said they served good food and that is where I wanted to be, in the hospital with the good food.

Had I realized just how long I was without food, I would have screamed louder and been angrier at those around me. I did not recognize night or day because of the darkness of the cubicle. The attention I was receiving was around the clock too, so there was no night time for sleep nor morning wake up time. Time was a void, and suffering became the measurable evaluation of my existence.

I wanted my family around all of the time because I thought I could persuade them to take me out of the hospital, or at least, put a stop to the horrible treatments the doctors were tormenting me with every few hours. I screamed at the top of my lungs, but nothing would stop them from treating me. The treatments were more painful than the fire itself and I began to think of the doctors as enemies who wanted to cause me more anguish.

It was a time when I demanded statements of continuity from those around me; I wanted to be reassured everything was normal and fine. I did not have the slightest concept of the total despair existing outside my little cubicle: the plans for a massive community funeral for those who did not survive, the anguish of parents keeping bedside vigils, praying for their children holding on so precariously to life. I was merely aware of what was happening around me; my sphere of interest had grown very small.

I remembered that on the evening of December 1, my father was scheduled to attend a large party for the Democratic Organization in the ward. This party was held every year and I always wanted him to tell me what they served at those fancy banquets. I especially liked hearing about the desserts that were works of art molded in sherberts, ice cream, and whipped cream and the fancy high-tiered cakes. I asked him if he had gone to the party and my father, still suffering from shock and anxiety, pretended he had attended the dinner. He stood at my bedside and made a large elaborate story about the fancy desserts, but as he explained the details of the mound of ice cream, he started to cry. I was very satisfied hearing his account of the party he did not attend, and I was happy in being reassured there had been no interruption in life.

Another thing kept bothering me. I had been wearing my father's cuff links the day of the fire. They were in the shape of little old fashioned alarm clocks and the hands of the clocks were set at three o'clock. I was so proud to wear them, but I lost them in the fire. I told my father about this loss, and told him I felt badly I did not know where they were. I have little scars on my wrist in the shape of round circles, caused by the cuff links becoming so hot the metal seared into my skin.

The mucilaginous leakage from my body sticking to the bed sheets was making me sick to my stomach. I could feel the oozing from my face and this liquid ran down the back of my neck and formed puddles in which I lay. The odor from the smoke still clung to my hair and I longed to have my face washed and a good shampoo.

At this time I did not know I had suffered a fractured skull from my fall, and that there were many little stitches holding my skull together. I did not know the offensive odor smelling like rotten meat was my own raw body being exposed to air.

Sick as I was, I kept longing to see someone who was in room 209 with me, and I also wanted to know who had died.

During this confusing time, I kept having difficulty recognizing visitors and the people taking care of me because of my short sightedness. This frightened me and added to all the bedlam I was in when I couldn't tell who was attending to me. There was one exception, an army nurse who wore the muted yellow nurse's uniform of the United States military. I liked her so much, perhaps because she alone in all the pandemonium represented order to me . . . and I could recognize her. She spoke to me and told me as much as she was allowed about what she was going to do to me, and then it didn't seem so bad. It was very important for me to be able to associate myself with order during this time and to know what was happening to me.

Even in the most traumatic moments, I found orderly routine brought me relief. I never saw this nurse with my glasses on, but I was comforted by seeing her distinctive pointed cap as she came in and out of the cubicle. I liked this nurse also because she told me stories about being in the army. She knew how to help me, and she usually just had to tell me what she was going to do. If someone didn't first tell, I screamed out loud, but sometimes, even if I knew what was happening, I would scream.

Once a doctor was going to test how far my eyes would open and he told me it would not hurt, which was not exactly true, or his definition of hurt was not mine. I remember him looking into my eyes with a bright little light and I told him if he got his hand near my mouth I would bite him, I was so hungry. He laughed and ordered me some ice chips to suck on. I don't recall how many days it took me to get food, but finally I did convince someone I was hungry. It was the nurse with the army cap who gave me my first liquid—it was only tea.

Since I was in the pediatric section of the hospital, I could see into other rooms across the hall because they all had windows. Right across the hall was a room full of little children in cribs and I could see their blurred figures jumping up and down in their beds. I did not see the protective nylon mesh covering that confined and kept them from falling out, so I became very nervous, and I told my mother to tell those little guys not to jump so high. My stomach did butterflies every time they jumped up and down.

Visits from the family were comforting, and my parents stayed day and night. My aunts and uncles came, but not my two sisters and my brother. My brother, Patrick, was only three and not allowed in the hospital. One sister was an airline stewardess and flying back to Chicago.

There was such a great to do about asking if my sister in the convent would be allowed to come and see me. She was just starting a cloister period that was to last a year and during this time she was not supposed to see her family. A Dominican monk had to look up the Canon Law to see if it was ever permissible for a novice to leave during her novitiate and then return to the convent. Somehow, it was arranged. My sister Dae was granted permission to come and see me.

My older sister LaVelle arrived first. I wanted my family around me because I hoped to cajole them into stopping the doctors who were treating me. There was no chance of this. No matter how much I wheedled, the treatment continued, and what was worse, the doctors requested members of the family leave the room, and they dutifully did.

Another eighth grade girl from classroom 211 was put into my room and she moaned like a dying cow; I was not very gracious to her and shouted for her to shut up. I certainly did not need any such distraction to my own pain, and I was becoming very capable of giving orders.

To add to my woes, I was in a bed far too short for me because I was taller than the average pediatric patient, and this made nursing care difficult for both the nurses and me. They constantly had to struggle with the problem of keeping me from slumping down and my feet from dangling out of the bars at the foot of the bed. Me feet were burdened with many tubes, cut downs, and even a shunt in one ankle to relieve the work of my kidneys because they stopped functioning some of the time.

My sister Dae came from Dubuque, Iowa, and I was so relieved to see her. She was a nurse's aide . . . and certainly she could get me what I wanted. She would protect me, she could see how I was being handled and hurt, she would come to my aid and make everyone let me alone and give me something to eat.

But she didn't. She stood back with the rest of the family and tried to comfort me with futile words of encouragement that did not stop the pain, and certainly did not stop the doctors from hurting me.

SUFFERING HAS NO TEACHER

I was pinned to my bed and I was alone. I was frightened so much of the time and I wanted to go home. I wanted to stop all of the sickness and I started to feel I could not endure any more, I could not go on. My family lovingly stood by, steadfastly trying to give me encouragement, but I knew how much my pain was making them suffer. I wanted to die, now. I only thought of dying as being an escape from excruciating pain, and not as actually leaving the world. Many times I thought I was being killed by pain.

My sister, Dae, was the only religious person I wanted to see during the first days of the crisis. She kept trying to soothe me by telling me how nice it was they had given her special permission to leave the convent and come to visit me. I was not impressed with their benevolence and reminded her I had jumped out of a window to get her to come to see me. I did not realize she was granted this special dispensation because the family feared I was dying.

Outside in the hallway one of the doctors gave my father the grave news. There was a possibility of my having serious brain damage and I might lose the use of my hands because they were so severely burned.

It is difficult for me to give an accurate account of the many emergency crises confronting a seriously burned patient. The awful pain is only secondary to what the entire body is undergoing.

When an extensive burn destroys large sections of skin (actually the largest organ of the body) this protective covering of the body must be replaced, and the patient constantly monitored for signs of shock, breathing difficulty, infection and kidney failure, as well as pneumonia. Burn patients lose protein in the fluids seeping from their bodies and this must be replaced, but as the intravenous feedings are pouring fluids into the body, other fluids are oozing out almost as fast.

Many times I felt comforted knowing the same treatment was happening to everybody around me. At least it seemed in many ways normal, and it almost made the pain more bearable. There was safety in numbers, and during this point in time, I started to believe I would find my friends as soon as I could leave my small cubicle.

There was some other place where my friends were, another happier place in the world. I did not know precisely where to place them—perhaps in another ward of this hospital, or another hospital, whatever place it was, I would find them. I never thought much about anyone having gotten out untouched. I kept thinking the entire building was destroyed by fire and everybody else was either missing or injured or dead.

Several weeks passed before I could conceive of anyone escaping or getting out alive and well. I never thought about the children I saw escape; I could only concentrate on their being injured or dead. Being alive was better than being dead, but was it?

I dwelt with lingering thoughts on death, so much so it started to be a source of comfort to be sought, especially when my body was touched by anything. Then I screamed. I had suffered enough, I could not stand more pain.

I remember the time they had to insert the dialysis shunt into my ankle to relieve my failing kidneys. It was a life-saving measure, but I was convinced they were cutting off my foot. It seemed no one would stop the horrible misdeed. In the burning and falling, the sirens, the inhuman screaming of the fire, I thought I had experienced all the rages of hell, and it terrified me no one would ever know where I had been and what I really was. I could never reveal to anyone that to me my experience marked me as an evil person. Many years later I discovered it is natural for a child to associate pain with punishment. But I took my guilty feelings with me far into young adulthood, believing I was an evil person being punished through my misfortunes.

CHAPTER XI

I did not realize how hard the work of surviving was going to be. I was never able to anticipate the medical crisis I had to go through next, but I could comprehend there would be pain. I constantly pleaded for food. If I wasn't complaining about the pain, I could divert myself by thinking about how hungry I was. I was mad at everyone around me for not feeding me — and I still thought there was a conspiracy to starve me.

Another thing bothering me was being in a bed that was so sticky and wet. I did not understand it was not the bed sheet that was dirty and wet, but it was the drainage from my own body. My body was a massive open sore.

It hurt to have the bed linen changed, the weight of a sheet was too heavy against my body. I begged my mother to make me more comfortable, but there was no way she could touch me without pulling the burns and hurting me all over again.

Days passed. I do not know what day it happened because I lost all sense of time, but at last there came a day when the doctors told my parents they were going to begin a change in treatment. I was to have open air therapy. This meant I would be transferred to a surgical floor and placed on a Stryker frame. On the Stryker frame only sterile gauze would touch my body. Private duty nurses around the clock would attend me.

The thought of being transferred did not excite me very much, because anything short of going home did not interest me. It was the thought of being moved and being touched that bothered me more than my next destination.

I could not move any part of my body and I didn't understand how it would be possible for anyone to pick me up when there was no place to touch me. Then, too, I was five feet six and felt very big and long, and I couldn't figure out how all of this was going to happen without my being dropped.

I didn't trust the people who were going to move me. In fact, I was so apprehensive the hospital staff said my sister, Dae, could come and help with the move and make sure I was not dropped. They pointed out, if I were dropped I could holler at my sister, or at least, make my mother holler at her.

Leaving my dark, little cubicle after so many days, where I had been carefully watched, monitored and cared for, would prove to be a big step towards my recovery. Soon I was on another floor, in a different room, and in a very precise manner the routine towards recovery was set up.

During the next months, I was to take an active part in my recovery and in doing so I learned to look with eagerness at my body while new skin grafts covered burn wounds and grew. I watched and became excited with the many changes I was experiencing. I did not know then, or come to understand until many years later, what a remarkable instrument the human body is.

The journey to the fourth floor proved one thing; I looked very bad. I did not mind when my family cried because they were part of me, and I never thought to ask why they were crying. I did not think about how I looked until outsiders came to see me. My hands were still bandaged like two big clubs and I was covered with a blanket. I never connected the weeping of my body with the pain I was in, nor did I grasp the rawness of my body as being connected with the pain. I certainly felt the pain, but I was too confused to make the connection.

SUFFERING HAS NO TEACHER

On the way down to the fourth floor some people standing in the hallway looked curiously at the bed and gasped at the sight of me. This frightened me. I forgot the terror of being moved and became upset at someone gasping in horror. There was a murmer in the crowd. "The McBride Girl, The McBride Girl."

I did not know what was happening, and anyway, we were moving along too fast for me to do anything about it. I was hurt, but this time it was a different type of hurting. It was not painful like the body burns, but hurt me inside and made me aware of how awful I looked. I guess I soon forgot what happened on the way to the elevator because in a few minutes I was involved with the nursing staff transferring me to a Stryker frame.

In all my worldliness of thirteen years, I already knew about a Stryker frame. It is a metal frame (6 ft. long, 18 in. wide) stretched with canvas backing. The patient lies on one frame and another frame is placed directly over the patient's body. Sheets are pinned around both frames (with the patient sandwiched in the middle), both frames are connected to a small circular gear that permits the patient to be flipped from stomach to back and visa versa.

Frequent turning of the patient is helpful in preventing bedsores on patients who are bedridden for a long time as we burns were.

I was told the Stryker frame I would be using was from Loretto Hospital. It had been used by a neighbor boy who had been in a serious car accident and was paralyzed from the neck down. I went to visit him when he was in the hospital, the year before, with other students from my seventh grade class. He was one of the strongest persons I ever knew lying in his Stryker frame. I often thought about him. There was every good prospect I would get well, but there was no hope for his recovery, since his spinal cord was broken.

I have never believed some people are placed on earth to gain sainthood by suffering great hardships. However, in this case, I would have to make the exception; he was a living saint. This boy was a great inspiration to me during those anguished days of recuperating from the fire and for many years afterwards. I was very proud to be able to have the use of his Stryker frame. I think if I had not seen him in such a contraption, I would have been much more afraid of it. While I knew what it looked like, I did not know how scary it was to be turned stomach to back every few hours.

To be so confined was unbearable, it reminded me how helpless I was, and it was maddening not to be able to move any part of my body. It was instant aging for me, and I cried that I was turning into an old lady at the age of thirteen. The gentle nurses assured me I wasn't, but then I cried that I was a baby when I had to be fed my meals. There was no happy medium for pleasing me during this time, and I was not exactly adjusting gracefully to the circumstances.

The transferring went smoothly. I was picked up by the sheets upon which I lay in bed and gently lifted and placed on the frame. I think it took four or more persons to move me. My sister was in the group, and it was quickly accomplished. Everyone kept assuring me I would feel better on the Stryker frame, but I was not convinced for many months how comfortable the frame was for a burn patient.

The big transfer to the Stryker frame did not really make me feel any more comfortable. I wanted something for the pain but they assured me there was nothing they could give me for the pain. I was under the general opinion the only thing I was given for pain was more pain. I told them so in no uncertain words. I tried to reason with myself that everyone had a pain reserve and I had not reached mine; I would never have more pain than I could accept, but most of the time this was too difficult to believe.

I was placed in a recovery room that had been turned into a ward for other injured children. Much to my surprise, a boy was in the room with me, and I did not think this was quite proper. Not only was it a boy, but it was the boy who was the captain of my row and who never would wait for me to finish my papers.

Politics may breed strange bedfellows, but fires are a close second. In all my anxiety and of all my classmates I longed to see, he was the last person. I wanted to share my griefs with a girlfriend, but not only did he and I share the same row in class, and rode the same ambulance to the hospital, but we had to share the same hospital room.

I also found there were two other girls sharing the room and at last, I was with my best friend, Helen. I was happy to be with her because now I was with someone who really knew what I had gone through. She stood next to me in Room 209 and she knew exactly what happened; so my searching for one of my fire companions was over. Helen and I would get well together just as we hung bulletin boards together, and went downtown together to shop for the nuns.

Another badly burned girl, April, shared our ward, but she was such a little girl, only a fifth grader. I could accept being burned and having those around me injured because we were big eighth graders, but I was very upset to see little children suffer too. Since there were students from other grades involved, I started to think the entire school had burned to the ground. This was not true.

CHAPTER XII

Burn patients become delirious and will pull out I.V.'s and catheters during a siege of delirium; we are so angry at being at the mercy of so many life saving trinkets. To be whole and well one day and have our health change so dramatically was like being a prisoner in our own body. I was given the best treatment, but this was no comfort to me and anger—deep, raging anger filled my heart to the bursting point. I was too young to realize there were many experiences ahead of me I would have to undergo, and no preparation in the world could have told me how to survive them. Anger and instinct, the two primitive urges for survival, took over and I began my way back. I still had to undergo a learning process that would become as natural as breathing was for other people. Each learning experience would be growth, and the growth would affect the rest of my life. In a sense, my life would become fuller because of the suffering I would learn to accept.

But all of the maturing and accepting of my fate was far in the future, and during those early days in the hospital, I felt only pain and anger. No one would talk to me. No one would tell me what I had to know. I had to mourn. However, I could not start my mourning till I knew who died, and I did not find out the total toll for many months. Those around me were frightened about my possibly dying, too, and were afraid the news would be too grave for me to accept.

I was not the only one who had come to terms with mourning. The whole parish of Our Lady of the Angels had to mourn the dead. While teams of doctors, nurses, and lab technicians labored continuously to help us, preparations were being made outside the hospital to bury the dead.

Parents called neighborhood funeral parlors to seek help. Some funeral homes already were too filled and other arrangements had to be made. Funeral parlors throughout the whole city would be necessary to provide a wake for each child. Some children would share visitation parlors as they shared playing games and going to school . . . sharing a normal part of childhood was now a disheartening fact.

Every family was assigned a priest to help with the arrangements. Burial costs would be paid by the archdioceses but this was little consolation to the families.

Fathers and mothers would go to the funeral homes asking the most "experienced" relatives to go with them. Men, who that morning were troubled by the simple problems of life, such as car trouble and leaky kitchen pipes, were now confronted with leading their families in burying one or more of their children.

Mothers would be confronted with the tasks of separating the child's legacy. In rooms and on closet shelves, toys and games were left behind. A red sweater thrown over a door knob, a pair of shoes, some books and dressers filled with clothing and all the memories of what might have been, if the child were still alive, had to be faced. Children's bedrooms stood dark and silent as people tried to avoid the loneliness of the day.

Parents were advised to be brave and not to cry. Many had to deny their feelings and hold back the tears that needed to be shed. The strong controlling agent was the belief in their religion. People were afraid to question the safety of the school, the efficiency of the fire department as it would concern their faith in God. Since their anger could not be vented it complicated the details of disaster. When people wanted answers as to the cause of the fire, they were cautioned not to question their religion because questioning the safety of a Catholic school building was denying your faith.

The ritual need of mourning was denied, and people were conditioned to stop grieving. People were very frightened. They had to learn how to cope with the unseen catastrophe, to mourn, to express anger, to accept loss. People thought they were being punished for such a horrible thing to happen. Such thinking is natural in disasters.

A stunned people went about the business of the day, as difficult as it was to bury the dead. People needed heroes and figures to look up to, they needed to be busy, they needed to be part of the suffering. A network of hierarchy had to be established.

The children who perished were classified at first, as being "the chosen," as closest to God. These children were angels. Some of those who survived felt extreme guilt and pondered why they lived. Were the children who died smarter, prettier, more talented? The flames that swept through the school did not discriminate, they claimed the lives of all children—the popular, the shy, the quick, the dull. The sadness is that they were children—alive, human, not angelic spirits.

The children who survived the six rooms were thought of as special and could comfort others. Mothers would sit in funeral parlors quizzing the classmates of the dead children, as if these children could relieve some of the furor. "Did the child suffer?" "Why didn't they get out and others did?" "Were they scared?" "Did they pray?" These children stood trying to comfort parents, but they did not know how to comfort themselves. They were black and blue from the struggle at the windows and felt no one could ever comprehend what actually happened that day. Everyone was so totally, desperately alone in trying to grieve.

Many children who endured the turmoil of the fire, but were able to go home that evening, suffered beyond the loss of friends.

They had to face the monumental task of examining their own mortality.

Some children who did survive became spoiled and came to be thought of as blessed and could do no wrong. Others were neglected as their parents felt pain too terrible to bear and closed themselves off to the needs of the surviving children, and could only dwell on the lost children. The degree of hierarchy goes from those who were injured, to children who were not attending classes that day, to children who were enrolled in the kindergarden held in a separate building a block away from the main school building. People started calling the children who survived, "fire victims."

Interestingly, there is also a similar degree of hierarchy among the survivors of Hiroshima. The Japanese found it necessary to formulate a new word for their language to accommodate the phenomenon of survivors. The survivors from Hiroshima, are called "Hibakusha," which literally means "explosion-affected persons," and conveys in a feeling a little more than merely having encountered the bomb, and a little less than having experienced physical injury from it. In the official definition, the category of hibakuske includes four groups of people considered to have had possible to significant amounts of radiation: those who at the time of the bomb were within the city limits then defined for Hiroshima, an area extending from the bomb's epicenter to a distance of 400 and in some places up to 500 meters; those who were not in the city at the time, but within fourteen days entered a designated area extending about 200 meters from the epicenter; those who were engaged in some form of aid to, or disposal of bomb victims at the various stations which were set up; and those who were in the uterus and whose mothers did not fit into any of the first three groups.

Each of those involved, from children to parents to firemen, were left to their own capabilities to carry on as if nothing happened. Forgetting did not make the pain go away, it just nurtured unexpressed feelings and many suffered from anxiety, guilt and depression. It was this immediate time when people should have been encouraged to show their feelings, not hide their sorrows and deny the need of mourning.

Repressed mourning was growing and growing but no help was forthcoming. In *Death and the Family,* Lily Pincus discusses the problems of repressed mourning, "We have seen how the loss through death of an important person strikes at the deepest root of human existence, recalls the experience of previous attachments and losses, and reactivates the pain of earlier bereavements, physical as well as psychological in nature. The emotionally deprived and threatened child may have learned to avoid the fear of abandonment and isolation by denying feelings and pain, and many thus have laid the foundation for defenses against feelings, the agony of the final bereavement through death. If life does not help him to resolve the anxieties which underlie these defenses, he will go on repressing, denying feelings—feelings of love as well as hate, those of joy as well as sorrow—and develop the image of an unfeeling, unemotional person."

These repressed feelings were carried out through the whole parish by the development of this hierarchy.

A grief striken city did its best to arrange public ceremonies for the comfort of the bereaved and to honor the dead. Flags throughout the city were lowered to half mast. A funeral Mass was said in Our Lady of the Angels Church for the three nuns who died in the fire. The Chancery Office of the Archdiocease made extraordinary efforts to arrange a White Mass for the dead—a Mass that is generally reserved for the funerals of Cardinals and Popes. This Mass was held in the Chicago Armory for many of the families who wanted to collectively honor the dead. Thousands of mourners attended by special invitation to pray for the dead. The Archbishop officiated, assisted by priests from Our Lady of the Angels. Religious leaders from across the nation came to offer blessings for the grieving families and the dead.

SUFFERING HAS NO TEACHER

A large plot of land was prepared at Queen of Heaven Cemetery for the families who decided to bury the children in a special gravesite for the disaster.

In some cases brothers, sisters and cousins shared wakes and funerals. Many parents wanted private funeral masses for their children, so churches all over the city opened their doors to the striken families. Masses and burials continued throughout the city for the following week. A monument was designed months later at Queen of Heaven Cemetary dedicated to all those who lost their lives because of the fire. All the names were inscribed, indicating the burial sites of each nun and child.

No doubt these tributes of faith helped some people carry on through the final ceremonies of the week. Friends traveled from funeral parlor to funeral parlor by bus or car and stood in long lines to pay their last respects. Then they called homes to apologize for not being able to go to the funerals or burials because they too were burying a loved one, or had the responsibilities of a hospitalized child.

My sister, LaVelle, tried to attend many of my friends' funerals. Many of the children had closed caskets. Many parents sent wishes to the hospitalized victims. There was so much sorrow and pain. Most of the families' traditions in these homes were based on a Western-European hertiage centered around Catholic dogma and ceremony. The people were able to cope by using established customs from European orgins. For days neighbors tried to be helpful and brought cooked food, sat with babies, and helped clean house. Finally it was time for all of the pageantry to come to an end. The excitement of seeing newspaper reporters in the area, and then reading their stories in the papers each evening, stopped. The rest of the city turned its attention to the approaching Christmas holiday. The Parish of Our Lady of the Angels was left alone to recover from the disaster as best it could. But the miseries of the fire lingered on. The surviving children hid in repressed mourning. The second, unspoken tragedy of Our Lady of the Angels began—the masking of feelings.

Something strange was happening; a dark pall was settling down on the parishioners of Our Lady of the Angels. They, who needed to mourn and give vent to their sorrow, anger and bewilderment, were being told to be brave. They were not to give in to weeping, but to accept the fact that the holocaust had been an act of God — their little dead children were the fortunate—the special ones—called to heaven by a lonely God who needed them for angels.

Our Lady of the Angels fire made headlines that shocked the very fiber of man's existence. It not only saddened a city, but the nation, and in fact, the entire world because little children were the victims of such an indescribable tragedy. Perhaps during that time people were jolted for a second or two into wondering about life and where it all leads, and when does the pain stop? Maybe they puzzled over what is the purpose of continuing to hope life would carry on when life was subject to such capricious strokes of fate.

Everyone in our parish was to be different from that life-death experience. We were the people from Our Lady of the Angels and most of us wondered, "Why us, Lord?" But at the same time, we were being instructed not to ask questions, to accept humbly the will of God. The miserable doubts about the fire lingered on and on, and thoughts about the lost little children had to be locked within.

Anxiously, the parishioners waited for some official word about the fire, but the subject of the fire was brushed aside just as the dead were being brushed aside by the prattling of religious dogma. The gaunt school building loomed over the neighborhood, a constant reminder of that unforgettable December day. There was some talk of rebuilding a school on the same site. The bereaved parents found they dreaded having the fire-gutted building demolished so soon, something of their children remained in the building, and they could not bear to have this taken away from them also.

SUFFERING HAS NO TEACHER

It was time for mourning, but mourning was denied to people slowly coming out of shock. It was not enough for them to be told their dead children were "safe in heaven," that they were "God's litle angels." They found no solace anywhere. They could no longer turn to the religious leaders on whom they had always relied, and many of the faithful began to feel uncomfortable with their church. They bitterly locked the world outside and withdrew into their sorrow. Many would never again know the healing of laughter, but would always wonder why they were seasoned out to bear such enduring pain and hardship. These people looked at each other and saw their families, friends and neighbors for the first time in such a revealing light it frightened them, and they cried, or screamed, or remained frozenly silent. They prayed and felt guilty for not bowing in humble acceptance to their fate. They tried to reach out to others they saw suffering who had lost a baby. They saw the strain and shared the hurt of the families with burn suffering children. They tried, oh how they tried, to rejoice with neighbors whose children were all home safe at night, but in their hearts they wondered, "Why my child?"

The fortunate and the damned gathered in groups and hoped the grief would go away, and somehow they would wake up one morning and find the holocaust was only a bad dream. "So much of living is a bad dream," they said, "but this is the worst."

Their sorrows were not private, their agonies were not singular, they were all caught in a web of tragedy and were forced to do the private act of burying their babies publicly—babies who would never again be held, babies who would never again know life.

The impossible answers to questions of why it happened would never be found, nevertheless they continued to pray for forgiveness for feeling their loss. Daily they quietly thanked God for sparing a child, but always pondered why another child was chosen for death. Hate and anger mingled with sorrow and seared their very souls.

They became strangely isolated in their misery, and found no condolence in the knowledge their agony was multiplied many times. Shyly and awkwardly they tried to comfort others but did not know how because they did not know how to comfort themselves. The old familiar, trite expressions of sympathy over bereavement did not seem to apply any more, but they kept on repeating them because they were too distraught to think of any other words. They kept trying to help, but there was so much pain, pain that did not let up, pain that would continue forever.

Sadly they realized, for the sake of other members of their families, the desolate parents had to continue living. The chill wind of hopelessness, that ran through Our Lady of the Angels Parish that winter night, continued to be strong. It blew bitterly and was not relieved with the coming of warm spring days. Ours became a parish divided among itself, and held together only by a common sense of loss. Some were able to abide by the admonitions of the church and humbly accepted their fate as the will of the Almighty, while others continued to ponder in their hearts and ask, "Why?" and they thought constantly about death.

Many turned to prayer and prayed their dead children were blessed and safe. They comforted each other by saying that the dead were truly blessed, they were angels in heaven. Still, the people were apprehensive, and needed comforting; the survivors needed to mourn their dead; the injured needed to be healed, but there was no surcease. Unbearable tragedy had taken control of the community.

CHAPTER XIII

The doctors decided to use the open air method for treating my burns. I was anxious to see my hands and glad the metal plates and white clubs that encased them were to be removed. The metal plates were wrapped with my hands to prevent the fingers from closing completely. Now they would go. Several nurses assisted the doctor in this procedure.

The shock of seeing my hands was grave. I did not know what to expect when the bandages were removed and neither did anybody else at my bedside. What once had been long fingered hands had changed into immovable, crusted, swollen, stinky appendages. My fingers, no longer slender and graceful, were in a paralyzed claw shape, and the colors were everything but natural. The skin had become the texture of gray decayed leather, like dried up leaves found in spring after all the color has been sucked away by the harsh winter. Blood gushed out whenever this gray, lifeless coating on my hands cracked open.

As ghastly as the unwrapping of the bandages was, and as upset as I was, something astounding was happening. For the first time I could remember I was in contact with people. I was being touched again. It was quite painful, and I really did not know who was performing the unveiling of my hands, but this started me thinking that I really could get well, just as the doctors promised I would. In all the misery I was in, I was being groomed for recovery, and I resolved to cooperate fully. For the first time, I was made very aware that the responsibility for recovery was mine.

The doctor finished his examination and left me with my hands lying on top of my chest now, still two big clubs, unable to grasp or to touch anything. The bandages were disposed of, but I was still unable to use them. My hands were dead to me; my hands were monsters; my hands were no longer part of me. The doctor's words, "Michele McBride, you will recover, you will use your hands again," was more godly than manly. It was like a huge puzzle or maze in which if I did this, then that would happen. I forced myself into thinking everything possible was being done, if I would only give it time.

Time was the one element I was not able to cope with. I wanted to get well now, not next week, so I tried everything possible to hurry up the growing of new skin. Time was a precious element and whenever I begged to go home and disregarded the conditions of my body, I was reminded firmly I would get well—it was possible—but it would take time.

I did not always believe this, but the hospital staff encouraged me to do the impossible, and that was to live. My constant longing to give up was always interrupted by my desire to live—spirit prevailed.

At no other time in my life would I ever be as disciplined or coached as I was then. It was a time when I gained valuable experience. It was also a holy time when I possessed warmth and love from comradeship, and that doesn't happen very often.

I was not always so philosophical. There were times when tears rolled down my face and I couldn't stop them. I cried so many times, silently, because of all the changes taking place in my body. Sometimes I did not know why I was crying. I could feel the moisture of tears rolling down my cheeks and I tried to blot them by catching the tears with my tongue so no one would know I was crying. Very often a nurse would come by and wipe my eyes and she understood I preferred no one say anything. Even though I was never afforded privacy, in my nakedness there remained some natural dignity and I wanted to be alone and not to speak, even perhaps to dream. It was during this time I realized good nursing cannot be taught, only felt.

It was during this time, too, my visitors were restricted to my parents only. This rule was enforced because of the constant threat of infection to the seriously burned. I did not like this arrangment at all, and I felt very isolated from my family. I was fine as long as they were around me, but after the first crisis days had passed, regular visiting hours were to be enforced. This meant I could see my parents for only two hours in the afternoon and one hour in the evening.

I felt stranded, and abused, because too much of the time there would be no one with me who knew me. It would be all medical staff, who just hurt and hurt as they treated me, and no one was there to protect me from the hurt. I was alone once again. Locked into my own woes, I was unaware that the wizardry of medicine was keeping me alive. The teams of doctors, with unfamiliar faces, worked around the clock prescribing medication, watching and caring for all of us. Shifts of nurses came from all over the city, and countless laboratory tests were run night and day to ensure we children were given the best medical attention.

I remember one medical crisis well. There simply was not enough of my unburned skin to give me shots and getting a shot on top of an area where there was a recent shot was painful. A nurse explained to me my thighs were already black and blue from receiving so many shots and I should brace myself for the next three shots she was about to administer. I realized every attempt to minimize the pain was a team effort all during my hospitalization. To insure this I had to be informed of everything that was going to happen to me.

They let me pick out and tell them the part of my body I wanted the injection, and that way I localized the pain onto an area I thought wasn't as sensitive as the others. The veins in my arms collapsed from the multitude of syringes of blood drawn for lab tests.

The advancement to the Stryker frame brought along another change for me. It required private duty nurses around the clock. This meant better nursing and it was a lot easier getting used to just a few nurses, certain hands touching me, instead of the treatment I had necessarily been receiving up to then. It had been difficult to explain to so many strangers the most comfortable and easiest way for them to treat me, how I liked to be fed, or even to hold a glass of water. For a patient who can't do these things for herself, there is an art even nurses have to master. I was lucky to get on well, and became friends with the nurses who spent four and a half months with me while I was in the hospital helping me progress.

The time had come when a new danger confronted me—infection. I was to be kept in an environment as sterile as humanly possible, and this meant everyone coming in and going out of my room had to wear a gown and mask. All bedding, bandages, and everything touching me had to be sterilized.

"Infection, infection, infection," became the code word for any reason why something could not be allowed that might hamper my body's fight toward recovery. I became very aware of this constant care to prevent infection, and often I thought of it as only making my life miserable.

The skin, I learned, is the natural body covering providing protection against disease. My skin was destroyed, so sterile gause fluffs and stinging, smarting antibiotics became my temporary protective covering. The worst prescription against contamination was the order no one was to touch or kiss me.

When my mother and I were alone, I begged her to kiss me because I could not endure not being loved. I refused to submit to being handled so cautiously and impersonally sterile. I was a "burn"—I was ugly, but I was still alive, and I had feelings, and my feelings cried out for tenderness. I was shocked someone could say no one should touch me. This had to be stopped. I wanted Mother to kiss me and I did not care how many germs it meant. I wanted someone to touch me. My sister, LaVelle, stroked my arms, one of the few places on my body that did not hurt, before the new ruling of no visitors except parents. I desperately needed the gentle stroking because I couldn't even rub myself to comfort me.

I thought I was losing energy and hope in the sterile medical environment, and grew depressed at being deprived of having visitors whom I did so want to see. I could not even have plants or flowers in my room. Flowers arrived every day, and because of the approaching Christmas season, they were beautifully decorated with Santa Clauses, candy canes, and snowmen. I was given one look at those floral arrangements from my doorway and then they were whisked off to the chapel, or put into the floral display at the end of the hallway.

During this time I was relieved when new glasses arrived from our family optometrist and at last I could see! Seeing is believing, and what I started to discover was how badly I was injured. The rack of pain I existed in was engulfing me, but the first sight of the wide expanses of black thickened skin and the oozing open wounds was another shock that almost overwhelmed me. I did not look human. I could not lift my head to see my legs, but I could see the ravages of fire to my body and it was ghastly. The darkened masses at the bottom of the Stryker frame were surely no part of me, and I wanted to disassociate myself from all that was happening around me.

I must not be reduced to one massive wound, I vowed. It was impossible this was happening and I was allowing it all to go on, but I had no choice, I had no authority, and I felt victimized by my own body. I was so sick, running high fevers, and I could not move or change anything around me. All my needs were attended to by others now; I was totally dependent. I could not pick up my head, I could not turn around, I could only permit all that was happening to pass. I was not certain I would ever be well again.

I was not only exhausted physically, but emotionally as well —

and I started to feel old; the aging process ran rampant. I would never again know boundless energy or fight off exhaustion. I was forced into accepting the fact I would never have a limber body. I would never again be able to rise out of a chair without first planning every movement, and I would never climb stairs effortlessly.

True, I could not voluntarily move my body, but I still had the capacity to reason and use my mind. I was becoming a victim in the space of my sick, destroyed body. To survive this ordeal, my mind would have to be my means of escaping my hampering body.

I resolved to learn how to go outside of my body to cope with pain. I would learn to find peace at levels where I never knew peace existed, but necessity would find it for me. I would learn to survive, and I accepted the bitter truth that every stage of survival would not necessarily be built upon the last stage. I would face every stage as a challenge in circumstances, not always agreeable to me, but I would sublimate with the inborn power given to people. I was soon to see pain does not kill, and pain cannot be shared, but must be endured alone.

Once when I had to swallow a foul tasting medicinal drink, one of the doctors offered to drink a glass of this horrible stuff with me. I asked why should two of us suffer drinking the glug? One glass was bad enough, let alone making another drink it, too. I forget how many pills I had to swallow between shots and I was never given one shot at a time. When they were to be administered, the injections were lined up on a tray in groups in the order they were to be given.

I also was to learn no pain is ever subtracted, but there is always the possiblility of pain being replaced. One pain does not disassociate a person from further pain.

I resolved to learn the art of suffering, not because I would never suffer again, but because I would learn to grow by widening my experiences, not lessening them because of suffering. I also learned my suffering made others very uncomfortable. For their sake as well as mine, it was better to try and meet pain quietly, because of the trauma brought to people around me. All this I learned when I faced the truth, my suffering was not going to be over in an hour or two. The pain would be with me the rest of my life.

The dreariness of the hospital room, with the dull wall coverings and the stainless steel pans and tables, was very dismal to a young girl who loved beauty. But something caught my eye, and I treasured it for the entire stay at the hosptial—a red blanket. It was not bright, apple red but rather the dull oxblood color of dyed wool so popular those days. I insisted the red blanket had to be on the outside of all the drapes over my Stryker frame. I viewed the blanket as magic and I would not let it out of my sight. I conjured up fanciful ideas about the blanket and it represented to me everything that was lovely. It was fresh and bright and not sterile. Its color was vital against the sterile sheets so dingy brown from having been sterilized so often.

That blanket was something I was not; it was whole, and it was pretty. I made such an issue out of the blanket all the children wanted one, but I would not give up my red blanket. I waited patiently until the nurse finished changing the dressings and draped a sterile sheet and then other blankets over the bar of the Stryker frame but I refused to rest until the outer covering was the red blanket. It was tucked around the bottom of the Stryker frame, and then the nurse turned the sterile sheet down and it hung over my shoulders.

My red blanket chased all the ugliness of the burns away, and it was fun. It did not resemble a hospital at all. Was I reverting back to childhood, or was I projecting into senility? Was I being foolish, selfish? I did not care. I was a burn, and the red blanket made me prettier. It hid all my ugliness and I refused to give it up, or even share it. I prized it, and made others prize it, because there was no other beauty around.

Man, they say, cannot live without beauty; I was no exception.

CHAPTER XIV

Having other smaller "burns" nearby reminded me I was a responsible eighth grader and I had to set an example and protect those young fifth graders. I attempted to do the impossible; I was trying to teach someone how to suffer.

I was very foolish so many times in my recovery. I wanted to set a good example of how to accept hurtful but beneficial treatment and how to suffer pain. I realize now everyone must find the one potential within himself, and no one can condition another on how it is done. There is no prescribed method . There must be a general rule on survival which is inborn in all people.

We "burns" lived in constant dread over the pain of the next change of our dressings. The amount of sterile gauze sticking to raw bodies and limbs must have been endless, and the work involved in changing dressings every three hours was limitless. The nurses first poured sterile water on our dressing to loosen the gauze fluff stuck to our bodies. This was of little comfort, because it still hurt when it finally came off. If the forceps slipped from the nurse's hand and jabbed our sore bodies, the wailing started. How exasperating it must have been to those tired nurses to have to listen to our begging them to stop, "No more please, leave me alone. I'd rather rot to death that suffer the pulling and the tearing of the gauze."

Antibiotics were then sprayed on the open sores and made the burning start over again. Every three hours all during the night and day, the uncomplaining nurses performed this treatment.

I was never given any sedation for pain because a burn patient can become so easily addicted to narcotics. I mastered the agony during the change of dressings by having someone hold my hands while I was biting down on wooden tongue depressors. I decided to stop screaming since the only effect the screaming had was to make my throat sore and hoarse. I was too alive to let the pain go unnoticed. I had to release it some way and this was the way I did it for many months. I often wonder how much strength I built up in my hands holding onto someone and squeezing their hands as hard as I could.

It is easier to accept pain, I began to reason, if one can find a purpose for the pain. With my religious training, I began offering up my pain for many reasons and causes. I imagined the worst suffering in the world and reminded myself that I was not suffering nearly that much. Sometimes I thought how dreadful it would be not to have freedom of speech and I thought that living in a communist country would be awful. Not to be able to talk would be worse than anything else in the world.

I had very little to rely on at the time, but I made very good use of what I had left. I thought how very good it was my feet were not burned. I thought of all the saints and martyrs I had heard about throughout my elementary school days, and I cheered myself by thinking at least I was not being eaten by lions, or flogged with stones. I could go on forever with terror stories of all the good Christians who went before me, and I kept saying over and over silently to myself "God gives us only the problems we can handle, and not one more. Everyone in the world has to suffer and must find his or her own special way to do it."

During this time, I asked the nurses and my mother to read to me over and over again the story of "Gabriel." This story was in the teenage section of the *Reader's Digest* collection on "Courage." It was about a little girl who died of leukemia at the age of nine. The nurses sat by my side during the wee hours of the morning reading about the little girl who died at Eastertide, and tears flowed down our cheeks until they asked if I really wanted to hear this sad story again and I said "yes" over and over. So we cried together.

I now think I was crying not so much for the litte girl Gabriel, but for all those suffering around me, and this suffering in the present made everyone so uncomfortable, we tried to avoid it. I memorized lines from this story that told about leaving your body when the pain got too bad, and I tried to do this—to leave my body. Perhaps I was meditating before it became popular and in vogue. It became a natural instinct when I tried not to think about the duties of the medical profession, and instead placed myself somewhere else in the universe.

I concentrated on corners of the room and pretended I could watch what was happening without really being there. It was strange that it was also very important I stayed in the room. I could not stop and leave the room completely because it would be dangerous. I thought it would be interesting to put myself up in a high corner of the ceiling and just let happen whatever would happen to the body down on the Stryker frame. I prayed to be given strength and fortitude for all types of emergencies, making it possible for me to go one more hour.

There were other times when I withdrew silently and wondered why I was forced to endure such agonies. Was this a punishment from God? What did I do to deserve such horrible pain? Then I stopped being morbid, and I remembered and started counting all the people I knew in the world who would share the pain with me. I listed all the relatives, on to the grocer, to every child I could remember who attended school with me, and I wished that my pain could be divided and five minutes taken by everyone I knew. Even if it were just five minutes, my time of suffering would be mercifully shortened. If only I could stop hurting and sleep, perhaps then relief would come and I could carry on.

I complained to my mother that if the next person who walked into the room asked me how I felt, I would tell him the truth. Any person who would come up to a hospital room and glibly ask "How are you feeling?" to a person who was burnt apart, deserves a snide retort. I was too tense to keep on answering "Fine, fine," and said I would rather not see people.

There was one person, a young newspaper reporter, who asked me how I felt and I replied, "Well done." I was tired of being polite. If I had been older, I am sure my vocabulary would have matched that of a truck driver, but the family kept reminding me I was a child and I had to do what I was told. I was sick and I was dying, and annoyed by people who had never endured burns, asking me how I felt.

Sometimes great waves of desperation washed over me, and I could not explain what was happening. Other times I was very aware of what was happening to me. There were other occasions when I did not even know what was going on around me. Not that I was hallucinating, I was just seething with anger and I wanted to be away somewhere and out of pain. I did not want the ordeal of growing new skin again. I would never move my hands nor have skin on my legs. I wanted to go home and have my mother take care of me. Worst of all, Christmas was coming and I was going to be in the hospital and I could not even have a Christmas tree. They might as well shoot me as to cheat me out of a Christmas tree. I heard the carolers in the hallways sing happy Yuletide songs and I was sad because I was not going home for the holidays.

CHAPTER XV

Late one night an old friend, a priest, came to visit me. He somehow was caught in the changing of the dressings but was not asked to leave the room. I am sure Father would have preferred being anywhere other than in my room at that moment of horror, even perhaps Saturday confessionals. Somehow he inherited the job of holding my hands while they changed the dressings. This was most unusual for hospital procedure, but they were shorthanded and I insisted someone hold my hands.

Later that evening he took the trouble to write me a letter I still treasure and which has encouraged me through the long years:

> December 14, 1958
>
> Dear Michele,
>
> First of all, I was talking to Montgomery (a French poodle) and he told me he sent you a card. Boy, is that ever a load off my mind! He's French, you know; indeed, he speaks with a broken accent but his tastes are not quite like yours and mine, for a while I was worried for fear he would send, well, snails. While we were talking, Gwendolyn (a cat) came along and acted quite huffy and puffy, and said that she sent you a card, too. Boy, you must be swimming in cards!
>
> Tonight, I was very proud of you. I saw how hard you tried to be brave and I saw you succeed.

SUFFERING HAS NO TEACHER

Believe me, Michele, bravery, like praying, takes effort—and you had what it took. God bless you for your good efforts. When I saw that, I wished that I could take your place or do something for you. But that was before I realized how much courage you had.

Keep up the effort. It isn't to please me, nor anyone else, but for the spirit of Our Blessed Mother of Calvary, and of Christ on the Cross. That's the sort of spirit the martyrs and saints manifested. God blesses you for it; His mother treasures you for it, and angels are delighted with you for it. I ask you to believe me when I say I left your room better for what I saw you do; and I beg you to face every trial, even the smaller ones with the same fortitude and courage. Even the Little Weasel had plenty (the fifth grade student) and why wouldn't she. She's got an example in you, thank God, and small and young though she is, she tries to imitate what she sees and hears. Being the light of the World was not easy for Christ and being the light of that room and under those circumstances is far from easy, and yet I know you are doing your part, and may God bless your good heart and a strong spirit.

If I get over your way, boy will I ever stop in! Please God, your recovery will be rapid and complete and then we can get down to brass tacks and plan what we can do. Surely, there are all sorts of things.

Now, it is very late, and I'm ready to keel over, so I'll say good-night to a very brave young lady whom I asked God and Our Blessed Mother to bless.

In Christ,
Reverend J. Heart

One of the greatest lessons I ever learned was gained from his letter. I learned that if you find qualities you admire in a person, don't wait to tell them. Do not be afraid. Take the time to sit down and write them, or just telephone, but tell them! We are not always given the advantages of time. To delay telling someone (not only that you love him, because that is almost always mutual) but that he is admirable and possesses many fine qualities, often is too late. We often neglect or delay doing this and time goes on and our friend is not there anymore, and we are regretful. I faced this in letting go of so many friends.

This letter from Father Heart has always been a reminder to me I must continue doing the best job I possibly can under the worst circumstances. Had he not taken the time to write me that night, I would not have known of his pride in me. Because of this unusual circumstance I gained much valuable encouragement and advice that has guided me through the rest of my life.

Father Heart's letters were always full of fun and always inspirational and because of them, I could always pray to God and receive strength to carry on.

Many times people commented about my being so brave, and I never liked hearing it because bravery was not something I had a choice over. The pain I had to endure was my life's job for those months. I was more often frightened than brave. I was frightened—so frightened about what was happening around me, and yet no one ever said it was okay to be afraid. It was such gentle prodding as given by Father Heart that made me want to get well.

About this time, the hospital doctors decided as many of the burns as possible should be put in semi-private rooms to insure the best treatment, so I was put in a room with April. She died a few days later. I would comfort myself and think when they changed my dressings at least she had to suffer no more. Death was escape from pain.

My next roommate was another girl who suffered a minor injury. She was lonely and wanted a playmate, but she took one look at me and started to scream. She was immediately transferred out of the room.

I was always aware I looked very bad, but I still did not know exactly how my face looked. I often asked my mother and the nurses for their compacts, but they all said, by a strange coincidence, they had dropped and broken their mirrors when they stepped off the hospital elevator. Once I was in a room where their was a mirror on the dresser and the nurse draped it with a sheet. I felt like telling her it was not necessary, since I had seen my reflection in the stainless steel basins and water pitchers that were around my Stryker frame.

I was not ready to confess to anybody I knew what I looked like; it would be difficult for all of us. I decided to concentrate on the hands and legs for a while and not worry about my face. I was more upset over what had happened to my hands and fingers. They looked so pitiful and non-human. I was scared I would never move them again, and I tried and tried to wiggle them constantly in and out of water.

An irrepressible urge to be able to take care of myself was the compelling force to get me to move my fingers. I could not stand being so confined. It was more than being restricted to a hospital, the real boundry was my own body. I became very aware of what body and soul meant, and very conscious of how we humans define our existence by our actions, by what we are able to do physically. Now I had to determine my existence without the aid of body language, and learn to live in a very different sphere of life. After losing part of my identity, my appearance and movement, still finding myself alive was a treat.

Space became important because I was so restricted. I strove to go beyond myself to find relief outside of my own existence. I could only rely on myself, and I found my reviving sense of humor a source of strength. I do not know how I managed, but very shortly after the fire I purposely incorporated fun and games into my treatment. This master plan took lots of doing. I was not very sure of those around me, but in the end, I had the best people near me actively participating in all sorts of fun and games. If they were not playful enough, I quickly chased them out of my games and my room. I drew on powers I never knew I possessed. No matter how difficult times were, I tried to laugh. What was more important was other people laughed with me. Because of the happiness growing out of so much laughter, some people took offense and criticized my behavior.

I was always aware some people thought I wasn't suffering in the proper way, while others admired me. How I enticed a sizeable network of the hospital staff into joining along with me in silly games, I never really tried to figure out. I wasn't in the best position for interviewing the people who were going to care for me, nevertheless I was able to surround myself with people I actually liked and grew to love. The small community of loving hands that aided in my recovery was a very select group of joking people. If they did not have a secret membership card to this club, I told them they couldn't belong.

Encouragement from many different sources came pouring into my hospital room every day, and many times, it was a briefly worded letter. Somehow, someway, letters were not thought to be carriers of infection that would do away with me. Maybe they thought the cancelled stamps had some sterilization powers behind them, but anyway, I could recieve mail. Many people took time to send lovely get well cards, and the steady flow of comedy and silliness coming into my room on funny cards made me laugh. My mother would open and read the cards to me, since I could not use my hands.

I could forget all the pains in the world and laugh at some of the letters I received, especially from a group of young men who were in advertising. Somehow they heard that I wanted to be the first woman president and they started a public relations firm to elect me president in 1980. These men took the idea and created one of the funniest campaigns that ever touched American politics. They were the founders of the T.C.C.O.T.M.F.P.C., which simply stood for The Chicago Chapter of the Michele for President Club. They were geniuses at entertaining me. Their letters made me not only smile but laugh so loudly many people wondered how I could be laughing so hard and be so seriously injured.

There were visits from doctors whom I did not particularly enjoy, and I did my best to hurry them out of the room. If those doctors did not regularly make daily rounds to check on me, I did not like seeing them and having them watch me. When visiting T.V. stars came to the hospital to visit the burn children, I pretended I was asleep so I would not have to see them. I was embarrassed about not only being naked, but without skin. The ordeal was terrible and I did not like it when people gasped at the sight of me. I still dreaded when people asked me how I felt. I was expected to say something mundane like "fine." How did they think I felt?

There was one time in particular my joking upset my friend, Helen. She was being moved in the hallway and I called out to her that we would not have to worry about dates for the eighth grade Christmas dance this year. Helen promptly told me to shut up. I did not realize what was happening to Helen at this time. I was confused about so many things.

To bring laughter into my life, in many ways I had to stop being a child. I had to enter the world of adults and understand what makes them laugh, so I resorted to the sharp, caustic quip. I could not be silly any more. I had to be responsible and disciplined. My parents always expected me to behave myself and follow orders no matter how difficult they were. Even so, my parents were an important part of my recovery and they tried to help me every day. Once when my mother came to visit me I was upset and she wanted to know why. I explained it was because I was supposed to eat my lunch lying on my stomach and I was expected to drink soup through a straw, but that day's soup had round noodles. There was no way I could drink that soup and I was disciplined and knew I had to eat everything on my meal tray. It seemed to me there was a master plan to make my life more difficult. That day it was pulling round noodles through a straw. Soon it would be walking on hot coals, just because it was good for me. I really felt this was the impossible task, and my mother saw to it I was excused from eating round noodles through a straw.

One morning a surprise visitor came to see me with his mother. This was a young man dressed in a dark suit, and he had very special permission to come and see me. He was the only stranger I heartily appoved seeing during my whole hospitalization. He was burned too, he told me, several years ago when he was a teenager, and now he was fine. I was amazed to see he was dressed and had cloth touching his body.

I do not recall saying much to him because his visit was brief, but essential for my recovery. Until that moment, I did not believe there was a chance for me to have skin again. Nevertheless right before me, was living proof telling me I would have skin and it would not hurt to have clothes touch me. All the doctors, nurses and family reassuring me did not count because they did not know what I was experiencing. I was convinced they did not have enough background knowledge to assure me I would ever get well.

The stranger, the young man, made me believe that if someone else had done it, I, too, would be able to cover my body with new skin. I was extremely impressed seeing a dressed post burn patient. I marveled he was not screaming in pain when his trousers rubbed against his body. He came, and he remained only a minute and left, but he helped me so much when I needed reassurance that I would be well. This brief visit from a stranger made up for all the unexpressed doubts of those around me, and I started to believe I would grow skin. There continued to be seasons of despair, when the days turned to weeks and then months, when I thought I would never get well.

It was drawing nearer to Christmas and my first skin graft operation was mentioned. I did not trust the doctors to put me to sleep while I was on the operating table. After all the pain I experienced having dressings pulled off every three hours, I was sure they doomed me to be awake during the operation. I was very worried. The first skin grafting operation was approaching and I was scared.

It was well known by my nurses and doctors I wanted to be president of the United States and that my hero was Adlai Stevenson. So all of the doctors tried humorously to assure me it was normal for politicians to be afraid of "grafting." I was not laughing this time, and I often suspected most of the doctors were Republicans on top of everything else.

It did not seem that taking skin from one part of the body and transplanting it to another part of the body was going to stop the pain and make me well. I was also desperately aware I had to get skin on my body, and to do so I must be willing to go through the apprehensions of unknown surgery.

I was worried about being back in the operating room with that big light shining down in my face. I wanted my morning nurse to come with me into the operating room, because I dreaded being surrounded by strangers who did not know how to handle me.

The first operation was sheduled for December 23 and then I knew definitely I would not be home for the holidays. I was upset over this and cried again and again.

The day my surgery was scheduled, my sister, LaVelle and my mother were allowed to come and visit me. They sat down and stroked my arm and told me what was happening outside. My sister was engaged and it was nice to hear of the wedding plans, but I was worried because I knew I would never get out of the hospital. She told me I was to be her maid of honor, and I would not get well and grow skin if I did not have my operation.

I remember being wheeled through the hallways and arriving on the sixth floor of the surgical suite and frantically hanging onto my private duty nurse at the doorway of the operating room, pleading with her to stay with me. She was not scrubbed for surgery and could not be allowed in, but she comforted me and promised she would wait for me downstairs.

I became frightened as I looked around the surgery room and saw the gleaming instruments that would cut me. I did not like the bright shining lights in my eyes. The scrub nurse asked me why I just did not close my eyes. I told them I was too nosy and had to look around, but I did not want to see anything that was going to cut me. So they covered my eyes with a towel. I heard a voice say, "Michele, if you stop talking you will go to sleep faster."

I had a very positive attitude toward operations after the first one; it was not a high happy feeling, but at least I did not dread being operated on. Perhaps my reasons were not the best, but they made surgery a better experience for me and for those concerned. The operations were long and usually lasted over five hours.

Skin grafts became routine and I looked forward to them because before the operation I was given morphine and I would be out of pain for a few hours. My body relaxed and I floated happily off to untroubled sleep. I did not have to worry about the world touching me. I eagerly looked forward to the hypo that lifted me out of the drudgery of living in an open wound. Even if it meant going through the pain of surgery, I was willing because of those minutes of relief. It was the only shot I welcomed and I could feel the numbness of the powerful drug lift me to the euphoric stages of contentment. The silliness of childhood returned for brief seconds and I could relax and be at peace with my body once again. Instead of dreading being held prisoner in my own form, I was dancing and laughing, and content in my body. I was happy, but I was high, too.

CHAPTER XVI

The rawness of my burns was replaced by the rawness of skin grafts. The donor sites, that were used to take skin for grafts, became new open sores. The large sheets of skin from those donor areas were shaved off in the thinnest, transparent, transplants and were cut in one inch preparations. These were then stitched on sterilized burn areas in hopes that those little islands of skin would take hold and grow together covering the open wounds.

As soon as the donor sites healed and grew new skin, that skin in turn would be removed and procedure repeated. The donor sites became new areas of smarting adding to the pain of my healing. Donor sites feel like a skinned knee stretched over a large portion of one's body.

I always fought to keep my stomach from being used as a donor site. I insisted it was the only part of my body that did not hurt and I wanted it kept that way. I was angry when the doctors told me we were short of skin from the usual donor sites, and my stomach had to be shaved in preparation for it to be used. They changed their minds while I was on the operating table, and to this day, I am grateful there is no scarring on my front torso from either burns or surgery.

The skin which is shaved off leaves a slight scar (in contrast to the keloid scar) and the outline of the long patches of removed skin never fades. The colorless layers of skin being removed makes the donor sites a permanent cicatrix of a healing wound, which is used over and over again, until the body is coated with a transparent covering. This is never actually capable of functioning as normal healthy layers of original, natural skin. The opaque squares of missing skin layers turn the donor sites a pale, anemic color compared to surrounding healthy, unmarred flesh. This stark contrast is always a constant reminder of the necessity to cover the body. Sometimes I think of it as a silent tribute to the skin that so often is taken for granted.

Growing new skin is a job, brutal and frightening, for the burn victim, until he sees the victory of accomplishing what at first seemed an impossibility—growing skin. I often thought it would have been possible for me to sprout wings and fly like a bird after I watched myself growing skin. In spite of my insisting vehemently every day, if not hourly, that I would never have a non-weeping body, that I would never be able to touch my body again without the burning, stinging hurt—it happened. In spite of the many fears, it did happen.

After each surgery, there was a rest from the grueling routine of having dressings changed every few hours. My burns were wrapped like a mummy and stayed bandaged for two days. During this time I was only turned every three hours and not a dressing changed. The luxury of not having my dressings pulled every few hours, also added to the pleasure of surgery.

The donor sites were covered with plain sheets of sterile rayon that would be exposed to the air. Large scabs would form, the donor site would heal every ten days, and then another operation would be scheduled.

However, during this time, a constant battle of itching went on. It is a sure sign of the healing process, and in spite of all the pain and other discomfort, every burn patient moans mostly about the constant itching. It is one thing to scratch yourself when you have skin, but it is impossible to stop an itch on an open wound. I always thought it was a great injustice to be bothered by a normal nuisance, and in this case an insurmountable nuisance. I fought with God because this was so unfair.

One doctor always assured me that pain is a good sign there is a brain left. I was tired and bored with practical jokes from heaven, but then I calmed down thinking at least I wasn't being thrown to the lions or stoned like the early martyrs. Itching meant I was healing. It was a good sign, but still what I would not have done to have been able to stop it.

Another thing constantly upsetting me was the smell of the burn. The nurses sprayed air freshener around the room, but this did not help. When my mother came, she gave me her hanky scented with the sweet perfume of "White Shoulders." I especially liked this after surgery to counteract the heavy odor of ether that mingled with my hair and the fresh bandages completely enclosing my body. I was always very sensitive about the putrid odor of my burns. I smelled like a sewer. I was oozing, I was reeking, but I was living.

Being near me took a great deal of getting used to. Because of the open air treatment, it was necessary to keep my room as warm as possible. So those who cared for me had to wear gowns and masks that were very warm in a room already too hot for everyone but me. The odors were overpowering and conditions unpleasant, but in spite of everything, there was an air of gaiety.

The hospital staff often engaged in games with me. The floor supervisor gave me a stuffed animal, a cat named Gracie. Stuffed Gracie and I had similar complications because she was red, and I thought that she too suffered from a severe burn. One of my nurses made Gracie a Stryker frame under the television stand so she could also be turned. Everytime I was given a blood transfusion Gracie received one too, red thread running through old I.V. tubing to look like the stuffed cat was getting blood. Once it was necessary for one of the surgeons to sew up Gracie's injured foot where the stuffing was coming out. However, Gracie fell into evil ways and bad habits, and developed complications I never had. Gracie often got hold of the ward's liquor bottle and became so drunk everyone reported seeing her in that condition all over the hospital. Either a resident, a doctor, a nurse or a janitor would stop by and report they had seen the drunken, red, stuffed cat stalking the grounds of the hospital. The more I recovered the more outlandish the stories became about Gracie, the alcoholic stuffed cat.

Every morning I invited two of the private duty nurses to come into my room and have tea. My own nurse came and the nurse from across the hallway since her charge was not a morning person and was better left alone. My tea parties were against hospital regulations, and tea cups could be found in many corners of my room, put there when the nurses thought a supervisor might be dropping by to see how I was doing. Another item sometimes found in my room was a resident trying to catch up on his sleep. Hiding tea cups was much easier than hiding a resident from a supervisor, but at least my room was always thought of as safe ground for the weary. They knew I would never tell, and they in return added to the madcap moments of my recovery.

The routine of changing dressings finally got to the point where it seemed normal and I learned to stretch out my mind to see how I could get rid of the pain. Finally they did give me two white pills for pain. They were aspirins. I was never able to convince the doctors I did not have a simple headache. Each and every day, I asked the doctors when I was going home, and each and every day they assured me that it would not be too long. Unhappily I saw Christmas come and go; then days and weeks went by. Again I was upset when I was in the hospital and cherry pie was being served for Washington's birthday.

I sometimes measured time by how long the treatments were. I learned not to ask how long something would take, but how many dressings were left to be pulled off. Minutes did not count anymore. The element of time I could measure was pain: the stabbing pain of the open wound, the excruciating pain of dressings being changed, and sharp constant pain of being burned. Anger flared and then went, and there really was no release for it.

Never being alone was hard for me. The lack of privacy and not being permitted the luxury of thrashing angrily about compounded the difficulty. I had to be self-controlled, behave nicely, and respect those who labored over me. I was permitted to scream in pain, get crabby, and even allowed to be sick sometimes. However, I had no outlet for, and did not know what to do with, the anger that encased my body as well as my encrusted dead skin. I did not learn about the anger for years, and I think I confused physical pain with the emotional anger, and they are very different. The consequences of pain I was able to master, but anger shamed me into feeling uncomfortable and inadequate.

There was so much exposure to the conflicts of life during the most vulnerable years of my adolesence that I developed unique impressions which were to color my whole adult life. I was not allowed the privacy of personal injuries, and I saw such dreadful suffering I often had to consider myself fortunate. I saw the increasing weariness of parents grow into hopelessness after many months of their child's hospitalization. I witnessed them struggling to keep their religious faith that was challenged on every side by despair. Many people who survived the consequences of the torrid schoolhouse fire and felt anger searing through them, had to keep silent because the holocaust was on the grounds of their church. For them to give voice to this anger would be a voiced denial of their faith. It was this fine line of belief that made it very difficult for me to express anger, because it would be confused with a denial of faith.

I was angry at the incompetent way the fire was handled, and knew in my heart it was caused by human error, and not a wish from God. I could never make myself clear on this point. I found my faith in God ever stronger, but I found the religion of my church unable to help me. Religion is human and not mystical as faith should be. I found I could always pray for support from God, but I could not always ask for support. Separating faith and religion became the untold hardship of this disaster. It was insurmountable for many who even now cannot distinguish between religion and faith. It was a time when everyone turned to religious leaders for solace only to find they, too, were caught in the disillusionment of this holocaust. All were stunned; all were suffering.

The awesomeness of dying frightened those around me, and we never mentioned it. Oh, I threatened that I wanted to die, but I really did not understand what dying was about. I just wanted to be relieved of the agonizing pain. There were times I was so frightened I thought I would lose my mind. Then there were life's aggravations I did not want to bother with. I did not like strangers coming into my room to stand and gasp in horror at the sight of me. I welcomed people who knew me, but they were not allowed to visit. Instead many religious persons were allowed, and I did not like making small talk with them.

Being young probably helped me heal and endure, but being a child was distressing when I had no authority to control the environment around me. I did not like people who had never been burned telling me they knew what it felt like, because there is no other pain in the world as terrible as a burn. They simply did not know what they were talking about. I did not like making allowances for them because they were so totally inept at being honest with me. People guarded themselves against my condition by pretending not to notice, and this made me very angry. It made me realize I was the only one who was going to pull myself through. At times there were so many emotions rushing through my mind, I did not know exactly what I was experiencing. I realized some experiences would prove to be valuable and fruitful, while others would prove to be destructive. All emotions demanded my attention and I was too ill and confused to sort them out—later I would have to resolve them.

There were all types of hassels, physical and mental. I became listless at being kept alone so much; I was worried I would not graduate; I tried to do homework, but I could not hold a pencil yet; it was impossible to write lying on my back or stomach. If I hung my hands down to write they turned black and cracked open. I often thought I was being asked to do the impossible. A teacher from the Chicago public school system came to tutor me, but the woman who was assigned to me was so affected by my condition she had to quit. My classes were suspended. I believe this was done through orders from the archbishop's office, and it must have taken an act of Congress to get such a privilege.

I was not upset about falling behind in my homework; I did become melancholy about my own condition and how I looked. It was a brutal realization to know the ravages of fire had permanently damaged my body and disfigured my face, and then to remember what I looked like before I was injured. Skin grafts, even though they were my own skin, were foreign to me. I always referred to them as "the doctors' work," but I took all the credit for the skin that I felt was growing back naturally. I did not like those little stitches holding the grafts. I felt disassociated from many parts of my body, but I could never speak about what was happening to me. It was a very strange sensation to feel distant from parts of my body for such a long, long time. I started to think of my body as a case, and I was not part of the body in that case any longer.

The body was not the Michele I knew. I could make her go away to other places. No matter what was happening to the body, I would still be safe hallucinating I was two separate identities—a body and a soul. No one would ever touch me again, no matter how terrible the pain was or bitter the medication. I was safe as long as I kept my soul and personality away from the horrors. I still retained the natural instinct of self-preservation against the hurt, but it was going to take lots of reconditioning for me to start thinking of myself as one whole person. The journey back to self would often be difficult.

CHAPTER XVII

I was healing, I was itching, but most importantly, I was eating. Burn patients lose so many vitamins and protein that it is essential they eat and often.

Burn victims usually lose interest in food because they are so critically ill. The usual burn patient loses about twenty percent of his normal weight, but I gained over twenty percent of mine. A doctor sat down one night and told me my skin would grow back only if I ate plenty. I was hungry all the time and I wanted to eat, so I didn't need much urging. I ate and ate and ate. There were hospital regulations requiring me to eat everything on the hospital tray in order to have food brought from home. So there I was eating three meals regularly, plus the home baked rolls my mother brought every afternoon. At night, my father came laden with either a pizza, or a hot dog and french fries for me. My parents' visits were like a kitchen on wheels with all the goodies they brought with them.

I drank gallons of liquid because I was told that would get me home faster. It was one of the few times in my life I could eat and drink anything I wanted and be praised for it. I feasted. It was one of the pleasures afforded me and I took advantage of it.

Getting some children to eat was a major problem, but it certainly was not mine, because it made common sense to me that eating would help replace my skin. During those months in the hospital I always thought it was my job to replace skin no matter how hard the circumstances. Another important fact about eating was that it made me feel I was contributing to my recovery.

One day I made a promise to myself. I would recover, I would grow skin, I would submit to any and all medical procedures. Every morning I would let syringes of blood be drawn until my veins collapsed, and at night blood transfusions poured it back. I would be medicated, fed, bathed, turned from front to back, have dressings changed, swallow countless pills, and I would live.

Almost every ten days I had a skin graft operation, and I lost count of the number. Some were successful, others were not. In spite of all the precautions, infections did win out many times and I ran high fevers, and then some of the grafts were lost. One time I became discouraged about so many operations and I decided I would not have any more.

The next operation went very poorly; I could not psych myself up for any more. Energy was running very low for everyone concerned, but somehow, someway, through all my fatigue, I summoned enough strength to push on with my campaign of recovery.

My hands were becoming flexible through constant exercise, and perhaps by my squeezing so many hands while dressings were changed. I became nauseated whenever I saw the layers of dead skin loosen and fall off in water, but it was interesting to watch the skin under the fingernails grow from black to white. It took months for my nails to grow completely free of burned skin but it was also a method of measuring my recovery. Many mornings during rounds I would wiggle my fingers to the doctor's delight. I prayed this progress would not stop, and my stiff claw-like fingers would fade in the dark as so many of the other miseries. I yearned to be able to feed myself and to brush my teeth.

After going pillowless for many weeks for my broken head bone to heal, I was finally given a pillow. Upon my request, the doctors always spoke of my fractured skull as a broken head bone because the terminology of fractured skull made me feel queasy. A broken head bone did not sound so frightfully dramatic.

My temperature continued to fluctuate, going from normal, to high, then lower. Many times I ran a very high fever, but I was always cold because of the wet dressings touching my body. I was still flipped from stomach to back every three hours and I was always on the opposite side from the one that the doctors wanted to see, and this annoyed me. I could not appreciate all the changes taking place when it meant I had to be turned on the Stryker frame.

Every so often another burned child was discharged from the hospital and I felt betrayed because I had to stay on. It was unfair. I was working so hard at getting well and I still had to stay.

Television helped me pass the time and I became addicted to the TV soap opera, and it is still an addiction I enjoy. I have watched "Love of Life" since I was ten years old.

During those long months of getting well something very special happened around me. It was more important than all the medication I received. I became beloved by the staff of the hospital. I do not know how to explain it, but in spite of all the suffering, all the losses, I was fortunate to be surrounded with people who encouraged me when my spirits were low, or when I was critically ill. They helped me find laughter and this laughter was infectious—the only infectious commodity allowed in my room.

This laughter flowed out of my room and spread through the hospital. I cannot explain this because it was magical, and no matter how sick I was, it never left me. From the head of the hospital to the chief of surgery, my loving, joyous friends helped me get well and restored me to my parents.

I will always remember how my father said every night I should be a good girl and listen to what the doctors and nurses said because their word was sacred.

I was not to question their authority for the treatments that would make me well, no matter what effort it took. When I was listless and terrified my family and hospital attendants constantly reminded me how fine it would be to get well. My parents faced terror each night when they said good-night to me, not knowing if they would see me alive the next day.

Some people told me how brave I was, but they were so wrong. I was not brave, I was obedient. The constant encouragement, made my obedience the key to my getting well.

Part Four

Helen Dies

CHAPTER XVIII

I watched you die through hallways and doors ajar. Our rooms were connected by an adjoining bathroom and I could peek at you while you slept. The softness of the night lamp shone in your room during the hours when our night nurses slipped away for the three a.m. coffee break. The glow of the light swept across the beige concrete floor, casting shadows on the wall, until your electric Stryker frame looked like two, huge, hula hoops hinged together, with you nestled in between. I could hear the silence of the hospital during the shadowy early morning hours when procedures were halted to the barest minimum for other sleeping patients, but not for us, because we were strapped to a continuous twenty-four hour shift of treatment. It was during this time I wanted to call out to you, but I was afraid my voice would rumble through the hallways, calling attention and disturbing the three a.m. silence. Usually only the muffled voice of a nurse's aide answering a patient's light, or the shrill squeak of the elevator's chain door broke the silence. I wanted to talk to you, to be near you, but the situation seemed too dangerous.

I think we both knew you were dying, and neither one of us wanted to admit it. I kept asking the nurses, and it always hurt to hear of your slow decline. I overheard the nurses telling the doctors on their morning rounds of your progress, and the terms they used made me mad at you. They said "your intake was not good, your fever was high," but the one that sticks out in my mind was your "uncooperative attitude." I was mad at you for this. Now I realize the dying just make the living so damn uncomfortable they are called "uncooperative."

I remember clearly the last time you spoke to me. During all those months in the hospital you and I caught little glimpses of each other, but never said anything to each other. Then one day you did speak to me. It was the second time I was forced to sit in a chair, and I was not happy with the novelty of getting up off the Stryker frame. My legs were stretched out before me all stiff and black, covered with tight alien skin and dotted with many open sores with streams of sticky red blood running down from them. My legs were so rigid and heavy, they were foreign to me. I could hardly believe they were actually any part of me, since I could not move them. I felt deceived. I did not want to relax and try to fit the contours of the chair. I was dizzy and desperately wanted to get back on the Stryker frame which had become an old, comfortable friend for me. During this time, Helen, you were being wheeled outside in the hallway and your nurse stopped you at my door. You looked like a cocoon all tied up in your electric Stryker frame with sheets and blankets wrapped about you and your head peeking out. Your face was so pink, and your hair was tied in pigtails.

You said "Hello," in a tiny voice but already we were indifferent to each other's needs — and it was so painful to see each other. I thought I would start crying because I did not want to be in that chair and I could not move to get back into the Stryker frame.

I was trapped, and it was more than being trapped in a chair. It was realizing I was trapped in a body which did not look pretty or work. I was so worried about falling I could not answer you. My legs were itching. I could feel the warm blood dripping down them and I could not bend down to wipe the blood. Pools of blood were starting to form puddles on the sterile sheets, making the scene even more bizarre.

I looked up at the doorway which you filled with the Stryker frame and my eyes were beginning to tear. I was holding tightly to the arms of the chair. You bent your head to one side and spoke to me in a low voice, saying, "I wish I could do that."

I did not say anything to you that day or ever after. What were we reduced to? Little girls who once shared every school day by passing notes, and now we could not speak to each other—both wishing for the pleasure of being capable of sitting in a chair.

The competition between us was starting to grow, my recovery and your decline. Before the fire we did everything together—pinning up bulletin boards, going downtown, and we never did things against each other. I liked you so very much. I never worried about who was the smartest, or prettiest, and I never stood in awe of you, but shared with you.

Now it was all changing, and I did not like you anymore. You were not my friend. I would train myself to think of you as the "uncooperative patient" who did not want to talk or laugh anymore. I even thought you wanted to die. As I progressed, you faded; as I joked, your rigidness became more apparent.

I was angry at you for dying, and this anger did not subside for years. Watching you struggle made me feel so inadequate that I soon joined the conspiracy of the living and avoided your dying. I, too, mastered the evasive manner society demands, and did not talk about your death in polite company. It always made me feel foolish and dumb pretending death was something that would go away if I avoided talking about it. Certainly the dying person would go away, but not death. Witnessing your failing more each day made it all so difficult. When I asked the nurses about you, they said you were resting comfortably. But I knew differently. I could hear the struggle you were going through. I could feel the sadness and discomfort of those around you, who waited long hours hoping you would not complain — or become aware of the savage red open sore which was your small body.

Those who daily cared for us must have wished that some way

we could slip outside of our bodies and not be there when we were treated. The constant begging of "Please don't hurt me," was wearing on everyone—patient, doctor, and nurse.

One day I heard a ghastly, inhuman, growl-like sound coming from you and I wondered what was being done to you. It frightened me so much that I wished you would die. Your taking so long to die made me feel uncomfortable, but it was not you that made me feel so badly. It was all of the rest of us pretending you had the choice to decide to die and when, that was making the stress between us. There I was, wishing a friend would die—what type of horrible person was I turning into? I must be the most evil person in the world to wish such a contemptible deed. So, as self-protection, I built a fantasy about how pleasant death must be. After all, so many others had died and escaped from suffering.

There was this constant conflict of interest; death in the hospital was defeat for those of us who survived, but death at the site of the fire was a blessed event. Those children were elevated to God's angels.

It was late afternoon, the time of day when light bulbs turn away evening darkness, and I was alone in my room. It was near the end of regular visiting hours, and the muted voices of departing visitors pushing the elevator buttons and talking about the condition of patients they had visited could be heard. These visitors bundled up in winter coats. They always carried brown paper bags stuffed with familiar treasured items like a pair of soiled pajamas that needed to be laundered. Usually it was like the changing of the guards. But today it was different. I could tell by the rushing footsteps and the spinning emergency cart wheels that there was danger on the floor — your death was imminent. The hospital's paging system called for doctors; student nurses ran down the hall; then I heard the shrill cry of your mother. You were dead!

I sat in bed, deserted now that you were gone. My body cringed in sorrow, and I didn't know what to do. I sat alone and kept hearing people as they gathered outside in the hallway. The familiar sound of a nun's rosary beads swaying against her habit mingled with the other sounds the business of death always requires.

Someone came in with a dinner tray and I said, "I know."

The evening news blurted the details to me. It was as unreal as watching a science fiction movie when a small black and white picture of a man announced on the TV screen that I just lost a good friend. The newsman was saying, "The Our Lady of the Angels School fire claimed its ninety-fourth victim at St. Anne's Hospital today. Miss Helen Anfes of 2095 Pine Grove died of injuries resulting from the fire that swept through the Catholic school December 1, 1958. She is survived by her parents and five brothers. A funeral Mass is planned for this Wednesday in Our Lady of the Angel's Church. There are six other children still hospitalized from the fire. The architectural firm of Dudley and Brown has been awarded the contract to construct a new school on the site of the old school building that was destroyed by fire, leaving three teachers and ninety students in its wake."

The little television told me what happened, but it did not tell me what to do. I knew I would constantly be assured everything was fine with Helen—she was safe in heaven, but somehow, knowing my best friend was dead was not okay with me. Anger, guilt, and aloneness swept over me. I knew this was supposed to be the time for me to be brave and not cry and to say dumb things like "she is safe" or "out of pain," but never once would I say, "Helen, I miss you." This would be too difficult for all those around me to hear.

All those weeks of our suffering together were over so suddenly. I thought I knew what being dead meant by reading the Mass card with the names of all the dead children printed on it, but when you died, I realized that list was as remote to me as being an angel. It was as meaningless to me as all the lists posted on the school bulletin board announcing winners of the spelling contests, or the highest Christmas card sellers, or the children who gave the most pennies for the missions across the seas. The names were the same, but the difference was the children on the Mass card were dead! No, they were someplace else, in another hospital, or at home. They were not dead. I kept on imagining they were just missing or hiding. I could find them as soon as I got out of the hospital, too. They were just away some place.

Helen, I carried the guilt of wishing you were dead longer than the years you lived. I was angry at the doctors and nurses for letting you die. Because you died, I did not know what to do with these intense feelings of mourning . . . so I shifted the blame from you to those who attended to you. I could not accept your death. There must be a reason, a fault within you that caused you to suffer so long and then die, or was I looking for reasons I had suffered so long—and lived. Did the good die young? Was it true that all the children who perished became saints, and if so, why not I? Was I too bad to become one of God's angels?

Your hospital room was cleared out within a few hours after you were taken away, and there was someone in the room cleaning it. Both our adjoining doors were open by this time, but the familiar Stryker frame was missing. The ceiling light was on and a cleaning man was mopping up. The swish of the mop sent cold, stale bubbles pouring all over the floor. A regular hospital bed was moved in, and soon a new patient arrived. I knew he was a prisoner, because a police guard stood outside the door.

That night, whenever I heard the squeak of the elevator door, I wondered if you were resting comfortably. I wanted to call out to you, and I did. I wailed and wailed and screamed and screamed until I made myself sick. My night nurse sat beside me and patted my hand, but it was not good enough. Then she started to pray for you and say her rosary. I wailed and screamed all the more until I was given a shot to quiet me. Now, they said, I was being uncooperative.

It was almost funny. After all those weeks of suffering and calling for relief, at last someone was tired of hearing my screams. Now they decided it was time to stop the hurt, or had I finally learned how to perfect my screaming? Yes, when the agonies became personal, I was eased out of my suffering; when the doctors and nurses could no longer tell me everything would be fine, I was given relief. It all became blurred, all a bad dream. By the time morning came, the sedation had worn off, and you were gone, I was back in pain and in the hospital. No longer could we all pretend you were not dying. It was over and the ground rules were changed to "you were safe in heaven," so they all said. I was bitter that after so much suffering, you gave in to death. I think by this time, I must have convinced myself people have power over dying. I could not accept there are simply no rules for dying.

The afternoon of your funeral, my mother came and told me all of the details. When she said Father Green offered your Mass, I remembered a conversation you and I had last spring about how we wanted to be buried. I remembered you wanted Father Michales, not Father Green, to say your Funeral Mass. I am sure if I had died, I would have wanted no less than the Pope to say a Mass for me! But it was at that instant I took the entire responsibility of making sure you would rest in eternal peace. I was going to be your press agent on earth and plan funeral arrangements for you, just as you promised the same for me—we were silly little girls, planning funeral arrangements.

I guess we started on this funeral kick the previous spring, when a salesman from the Catholic cemetery came during a Sunday service and made a pitch for the new mausoleum. He left literature in the church and we took one of the pamphlets and examined the wonderful advantages of being sealed in a wall instead of being placed in the ground. I insisted I wanted to be comfortably placed in the ground, but you said it would be too cold; you wanted to be inside where it was warm.

It was that day you told me you wanted Father Michales to say your funeral Mass because you enjoyed going to confession to him and you were even thinking about becoming a nun. I was quite impressed with your decision about being placed in the wall, but becoming a nun, I wasn't sure about that. I already had a sister who was studying to be a nun and I did not think it was special. We did promise to carry out each other's wishes, but less than a year later, I was hearing about your funeral without your favorite priest. I did not even inquire where they buried you.

There were lots of rough times in the hospital, and I did try to behave myself some of the time, but no matter how hard I might have tried, I could not have gotten you into a mausoleum. This idea haunted me for years, but so many things haunted me for years.

Today I am grown and cautious about never volunteering to carry out anybody's last will and testament. How much time and energy I spent worrying about your final resting place I do not care to count. Nevertheless, dear friend, you are the one person I have come across in this life who taught me a most important lesson, and that is, how I would like to die.

Thank you so much, dear friend Helen.

Part Five

Home Came A Stranger

CHAPTER XIX

Easter was coming and there was talk of my going home for the holiday. Finally the doctor said I could leave the hospital for a brief visit. I had yearned to go home during those long months, and at last the time came.

I was to go home on Saturday afternoon. I would celebrate Easter with my family, and return on Monday morning. All too short a time, but I would be able to be in my home, see my baby brother, my dog Daisy, and my girlfriend who lived across the street and attended another school.

I set my hair with metal rollers the night before I was to go home. One side of my hair line had been shaved for surgery and was growing out in stiff, short ends, and there was what I considered a huge bald spot on the very top of my head where it had been burned.

Getting those rollers set properly was a problem since my hair was many varying lengths where it had been whacked off when it became matted and tangled in the firesinged parts.

The nurses kept insisting I should let them even off the jagged length, but I protested and refused to let them cut my hair. I felt more comfortable with the different lengths hanging around my face. Besides, I was determined I was not going to have any more of my body cut than was necessary. I did not want any more of me to be disturbed or altered by sharp blades—I wanted some part of me to remain intact. So many massive parts of my body would have to grow again, and it would take many years for me to become not only adjusted to a new skin, but for me to grow back into being Michele.

It was very difficult to get dressed. I'd been stretched out in my hospital Stryker frame stark naked for such a long time. Then I was accustomed to only a loose fitting hospital gown and I had forgotten the discomfort of tight fitting clothes.

I simply could not manage dressing by myself and had to call for help. At first I thought this was caused by my excitement, but then I realized I could not bend my legs or lift my arms. After all my directions telling people how to turn me and which way it was best to feed me, I now faced having to think about ways to tell them how to help me get dressed.

Since I still had open sores on my legs, it became necessary for me to evolve a way to ease into my underpants and keep the elastic band from touching the burnt places. My arms had to be lifted by my nurse to get my slip on and the tiny straps felt like armor against my shoulders. My worst troubles were the bobby socks. There was a large blood blister on my heel and the rubbing of those thick cotton socks was irritating, and the bulky cuffs tormented my sore legs.

All the layers and textures of my clothing rubbed against my body and made the sensitivity of my burns more pronounced. It was an effort to let the material touch my skin, and an effort to be aware of the touch. I had to wear house slippers instead of shoes because of the blood blister on my heel. I was not prepared to make all those concessions to my injuries. It was so peculiar to feel cloth hanging on me, and so peculiar to have to learn how to wear that cloth again. Every step in my recovery was marked by a new sensation I had never dreamed existed. The basics of life became a series of adventures, and I found new opportunities for learning in the smallest human acts.

My father came in the car to pick me up. As we drove along

toward home he turned on the radio and we heard the report that Michele McBride, Our Lady of the Angels fire victim, would be home for Easter!

It was a pleasant day and I saw that some of the winter still lingered on in just a few places, but little, bright tulip blossoms were peeking their heads out through the remaining dingy clumps of snow to welcome spring.

I wanted to be driven by the school. I had heard it was torn down. The only thing remaining was the huge foundation and of course the builder's temporary guard rails constructed out of various doors and used lumber. I wanted to see if there were any visible clues as to why it all happened, but there was nothing left. The school was gone and with it so many memories, however, unresolved questions remained.

I went home. After many months of begging and pleading to go home, it happened. I was finally home!

I needed help getting out of the car. My father had to pick up my legs and place them on the ground. The cold spring air felt good; it had been so long since I was able to enjoy the wind tickling my face. For many months I had been kept in a sterile environment, now germs were dancing around me and not threatening my existence. However, it would still be a long time before I finally was beyond the danger of infection and could freely enjoy nature again.

There were two cement steps into the house which I did not think about all the time I was in the hospital, but they were waiting for me to climb — and they looked mountainous. I struggled up them somehow.

To this day, I try to avoid steps, but I discovered the first day out of the hospital they would become a big part of life, and could not be disregarded.

There had been a combination of fear and excitement in my mind about going home. Every move, I knew, would have to be calculated so as to cause the least pain, and yet I wanted to be home so desperately that certain discomfort did not matter.

HOME CAME A STRANGER

Sometimes, I almost dreaded even the thought of going home. All the time I was in the hospital, I had thought of being no place but home, but I had always pictured myself at home as being well, not injured. I did not anticipate there would be so many difficult decisions to be made, and that going home would prove to be just an extension of the many compromises to life already made on my part. Most important, I realized I wasn't ready for all the adjustments I would be forced to make.

There were many people I wanted to see on this brief visit, but I could tell the sight of me was very upsetting to them. Although they had been told over and over how seriously I had been burned, they were not prepared for my appearance.

I think they all expected a perfect little girl again, because the newspaper accounts of my progress were always so optimistic and glowing. They were counting on seeing the same little girl who went to school on December 1, 1958, but what they saw was a thirteen year old girl-child who had been forced into the mannerisms of a crippled old lady. I was dismantled beyond belief and a horror to see.

There was an almost unbearable intensity of emotion in the living room that day—happiness that I was alive, and sadness that I was maimed and suffering terribly. The brutal fire seared me again; I was emotionally scarred even more deeply than I was by the holocaust.

The tension in the living room was intensified by the look on the faces of Chicago newspaper men reacting to my disfigurement. They tried to cover this by asking me questions. The excitement of having reporters there did seem to divert friends from my condition and lessened the tension in the room. I was nervously aware of how everybody felt and I was learning a bitter lesson.

Many times in life people are afraid to say what they are feeling, or perhaps they do not know what to say. They have been conditioned to hide their feelings. They do not realize that many times the silence hurts just as much as saying the inappropriate thing, and fear makes them silent.

During all this convention-like atmosphere, with people struggling to hide the shock of seeing me, man's best friend came to our rescue.

I was sitting in a chair with my legs uncomfortably stretched out in front, trying to balance myself, all the while hoping I would not slip onto the floor, when the most wonderful thing happened. My dog Daisy, who had not seen me for many months jumped onto my lap, reached up, and started to lick my face. She was the only one in the room who did not notice the ugly scars covering my body. Daisy saw nothing different about me and she sat on my lap and I held her close.

It felt good to be home again.It was Easter time,it was spring time. It was a time for rebirth, it was a time for rejoicing. A photographer snapped my picture and it made the front of the Chicago Sunday papers for Easter. Everyone could see how happy I was to have my dog welcome me and not be afraid.

The black and white newspaper picture of my dog and me made me very aware of the contrast between the scar tissue on my hands next to the soft naturalness of my arms. I plainly saw how two-toned I was, and for the first time fully understood how startling the sight of me was to others. Many people sent me the pictures from the newspapers, but I tried not to look at them. The scarring showed plainly and it shocked even me.

During this time, I walked with a cane. In the hospital it seemed like such an improvement after not walking for many months, but at home it was out of place. All those months my dreams of going home were of being home like I had been before I was burned. I never once thought of going home and being disabled. I forgot that the most important part of going home was going home burned. Home was the same as always, but I was not. My short stay that weekend gave me a bittersweet promise of recovery that would come. Even though I was in familiar surroundings, I was a stranger to everything around me.

No one else can ever really know the anguish a post burn victim goes through. What might seem a simple move for a non-injured person, becomes a challenge for a post burn person when the joints are so burned they cannot be moved without pain. In the beginning of my recovery, I noticed my pain caused anguish in the person helping me. Then I mentally resolved to chart every possible way of moving my body so the hurt would stem from my own body actions and not be passed onto others. I would absorb the pain, then I could not blame someone for my agony. Often I had to rely on others to move my body, and I would cringe in my pain and theirs and feel so badly.

I am surprised that my head hasn't fallen off with the many turns of anguish I twisted it in. Why was it one leg could be lifted off the bed with hardly any pain, and I would have to brace for the hurt when the other was merely two inches off the ground? Why would one keloid scar feel dead and the one right next to it twitch with an itching sensation? These are questions no one can answer because scarring belongs to the victim alone. The tenderness of those scars, combined with the emerging insecurities of my adolescence, added to my discomfort and unhappiness.

Getting in and out of a chair was work. I learned to use my arm muscles to lift myself off chairs, but it irked me to need assistance. When I think of all the energy it took to lift myself out of a chair, I wonder now how I did it. I could not sit at the kitchen table because I could not twist myself into the right position. How carefree life really was when I was a child; how carefilled life turned when I became a burnt child.

It was a happy first visit, but I was extremely nervous and self-conscious about having to do all the calculating over positioning my body, and I constantly feared something would bump my stiffened legs. I remember one time my little brother was running by me; I screamed at him to stay away from me. I was afraid he might stumble and topple over me, and I had no way of protecting myself from injury. I was upset for yelling at my brother, but I was mostly disturbed at myself for not being able to move away from his eager steps.

It was a good time. It was also a time that emphasized I was, and would always be, different. In the hospital it was fine to be a patient, but I did not want any limitations placed upon me at home. Home was supposed to be immediate and complete freedom. The longing and the struggle were supposed to be gone; the fears and anxiety I carried were supposed to be dissolved. It was not until a family dinner when a few friends came by to see me that I realized I was exhausted.

In many ways my first visit home was the initial step toward my recovery, but there were to be many years of work ahead of me before I would be well again. Every step to my recovery would be marked by some new experience, some new sensation I never knew existed before my burn. I must turn every challenge, every setback, every humiliation and hurt into an adventure, a new opportunity, a challenge to learn something from even the smallest act. The long journey into self would take years, years before I would feel anything but pain—little more than pain.

I was tired. I was weary pretending not to notice the shock of people, and I was tired because of my general condition. I look back today and believe that since the fire, I have never been anything but tired.

I did not mind going back to the hospital. It was going home again that held so many new fears.

CHAPTER XX

Strangely enough, I was almost relieved to be returned to the safe routine of the hospital. There was news waiting for me. There was to be a change in my treatment and for the next few weeks, I would require just the services of my morning nurse. After consultation, the doctors had decided to use silver nitrate to dry up the seeping, open sores on my legs.

By this time, I had become leery of any new treatment and protested, until one of the doctors assured me the new medication would not be painful because all the skin on my legs was dead tissue. When I felt the fiery sting of the silver nitrate on my legs, I was surprised to report the condition of the skin was very much alive. Sometimes I wondered how all this treatment was devised—who figured out when it was time to leave one stage and go into another stage. I never knew what to expect.

I know now being a child helped because I didn't know what I should anticipate, and that was very nice. Oh, I worried lots and I was afraid of many things, but I did not have the complication of being an adult and worrying about unforeseen problems. My greatest worry was that I would never be able to wear nylons again, and that my hair would never grow back. I did not worry about stiffened joints, poor circulation, and all the other physical and emotional problems that eventually became my mode of life.

HOME CAME A STRANGER

I was always treated with optimism, and never allowed to feel sorry for myself. I was instilled with a sense of my own responsibility for getting well; it was expected of me and there was no questioning that responsibility. The nurses and doctors did their part, but the ultimate result was up to me, and I was to do everything I was told.

A serious burn is the worst injury the body can undergo, with the odds always against the victim. The long hours of apparent recovery may be snipped away, and without warning the patient may suddenly die. Despite the deadliness of the long statistics stacked against me, I was given the will to endure, to go on, to fight, not try just one more time, but withstand the worst pain again and again, and to continue fighting for life. It was a time when life and death rushed before me. Many times I was caught in the entanglement of dying, of questioning my existence, and examining the qualities of suffering. It was a time of great faith and my prayers to God for strength were answered in so many ways I sometimes was amazed at the power of prayer.

I learned what to pray for, and not to waste prayers on silly things ever. I limited my prayer for special purposes. Learning how to offer my pains to God for the benefit of others motivated me to pull through.

I am still very self-conscious when people talk to me of my bravery, because they do not know how many people encouraged me to endure pain, and they do not know that I really had no other alternative. With so many people working with me to get well, my cooperation was almost natural. When all my energies ran low, when pain became a way of life, when medication had to be taken, one more tube of blood had to be drawn, support came to me from the hospital staff who were all dedicated and geared to helping me. These people were dedicated to the restoration of life. At the time I could not appreciate this, but I grew to understand it, and to appreciate that I had truly the finest medical staff attending to me. I was encouraged to grow back skin, and with every little advancement, I was praised and made to understand what a great personal accomplishment this was. I was favored by these people who coaxed me to go on. Some might say I developed a method of recovery that could be tagged as "inner strength," but I was just doing what I was supposed to do.

When I think back on the horridness of pain, I cannot remember it. I cannot recall the feelings of the ugliness of a sore body, but I have not forgotten the merriment that made me get well. As I progressed the medical staff became so festive and so lavish in their praise, I often grew downright embarrassed and ill at ease with my recovery.

Being tagged brave was for old soldiers, and I still wanted to be a little girl. It was hard for me to understand why I was getting well, when many others were losing their lives. This frightened me more than the fire, and the panicking crowd in the schoolroom. My jubilation over recovering was mixed with a feeling of guilt that I was surviving a disaster that had killed other children. I became depressed. I was angry at myself for feelings of happiness and considered myself morally imperfect. I was a teenager and relentless in self-criticism.

I celebrated my fourteenth birthday in the hospital and it was a double celebration with Gary, a classmate of mine who was also a patient. A bakery made a huge cake, half pink and half blue, and it looked more like a giant cake for a baby shower than a birthday cake for teenagers. It was funny to have our birthdays in hospital rooms and announced over television. He was still in isolation, but I was allowed to visit him, a big break for both of us.

The impact of seeing this boy, almost totally burned, and yet so uncomplaining, made a great impression upon me as it did on everyone else. I have always felt fortunate to have known him. I could see the outline of his body, but there was no skin on it. His face was beautifully intact as his spirit must have been.

In the eighth grade, we were not in the same room, but the year before we had been, and I remembered him as a boy who was always pleasant and a favorite with the nuns. He always had his homework done and was cheerfully willing to serve at Mass. He was athletic and excelled in all sports, but there he lay in his hospital bed, destroyed. Strange, the things you remember about people. This child was certainly one of the boys who gave promise of being a fine, handsome young man. I recalled that he often wore a green plaid shirt, and he was always friendly and even talked to the girls.

As I sat in the wheelchair by his bed, it seemed so long ago when we were just schoolmates, and now we were survivors. How suddenly life changes. What was happening to my eighth grade class, where were the children with whom I stood in line and giggled, the children I competed with in spelling bees? It seemed as though a million years were separating us.

I always heard nothing but the best said about Gary, and as he lay there in his hospital bed, I pictured him coming into the seventh grade classroom back door after he served a morning Mass. It was difficult to remain and watch him suffer so uncomplainingly.

Wishing each other a happy birthday was a great accomplishment for both of us that day. Gary never recovered. It was his last birthday.

How the medical profession has the will to go on and treat burns, and then in many cases, lose a patient after months of treatment, is incredible. How unselfish to labor over a burn patient against great odds—odds that seldom are in the patient's favor. Burn specialists must have the nature of card sharks and be big gamblers with the chances they take in investing in a burn patient.

I was coaxed into recovery and that is what I remember most about the hospital. The wonderful spirit prevailed even when I was most depressed. I believe the difficult part of my recovery was the depression. I could not see the depression, I could not measure it, but it was constantly upon me and as grave as the seriousness of my burns.

There were times in the hospital when I could not keep all that was happening to me separated and I would wonder why it happened. The implausibility of going to school and having it burn faded out of reality many times. I tried to deny the fire, and that I would never be reacquainted with my classmates. I refused to believe we would never be whole again. I found myself reaching out for information that would answer the questions of why it happened. Why was I alive? Why must I continue living? These thoughts kept coming in and out of my head. I could handle the physical pain so much easier because it was tangible, but what was going on in my head frightened me. I never knew how terribly angry I was. The anger was not acknowledged. It was turned inward. Because of this, it later had to be treated for many years. All my screaming because of pain did not release the anger I was experiencing. I felt guilty because I was so angry, and as anger seethed within, my anxiety increased, and my depression deepened.

There were also times when people could really hurt me, too—more deeply than the burns. I could deal with people who were experienced in the treatment of burns and brushed aside my appearance because they caught signs of improvement. However, there were lots of people who winced in shock at the sight of me and reminded me constantly of my scarred looks.

Once I was in the hospital hallway trying to walk. I was pushing a chair with my mother on one side and the floor supervisor on the other. A man, visiting another patient, came up and said in a loud voice, "That will teach you not to play with matches again."

That was but the first of many such cruel statements I would encounter. All of the kindness and words of encouragement were wiped away by such thoughtless statements, and I had to learn to deal with them. That was another type of pain I had to grow to accept.

The end of the long medical campaign was nearing, and the time was approaching when I could go home permanently. This brought many fears and problems, too. How would people react to my scars? The excitement of growing new skin would not be understood when the open sores on my legs were in plain sight. The agileness of my fingers would not be seen when I could not walk without being noticed. The exaggerated movements of my body would have to be minimized. I would have to cover up and conceal so much I would be weary before the day started. The groping for reasons, "Why did I live and others die?" continued to haunt me. There were times I became listless and withdrawn.

The difficulty I was experiencing in learning how to walk seemed endless and I was still practicing walking and climbing steps. I dreaded stairs. The constant coaxing from my nurses kept me moving whether I liked it or not. I was developing my own shuffling gait.

Going home was difficult, not only because of my limitation, but because I was going to a grieving neighborhood, where something very solemn remained in the air. It was a time for seasons to change, a time when the weariness of winter should have been turning into spring — a time when people tired of the long, cold, winter months ought to have been welcoming the warmth of spring. It was the time school children started to wear light coats and jackets as they skipped toward school. This year, the heaviness of the winter months stayed. Spring was not joyously welcomed; it came unnoticed.

Blocks and blocks of people remained startled, frightened, angry and repressed. They were confused. The emotional exhaustion of the fire had increased. Children who were not injured in the fire had to walk home from school and pass homes where children were missing. Brothers and sisters returned to homes where other children were dead. They had to watch their mothers hold back tears. They saw their fathers wordlessly slip out of the house every morning and silently come home, not knowing how to find the words to comfort their wives and living children. People remained silent, or burst out in rage to find some release for pent up emotion, until live children became subjects of this anguish. The balm of expressed grief was not sanctioned; relief must come through prayer, so the people were instructed by the church leaders.

Life dragged on, but there were many unresolved questions. Neighbors shunned each other on the street and a general malaise of depression covered the entire district. Children were shifted to three different schools, and were crowded into classrooms high on third floors of public elementary schools. Nuns continued to teach their classes as usual but many classes were broken up with the injured and the dead missing. Some students became hysterical and were sent home with notes instructing parents to do something about their children. Several times a false fire alarm sounded. These children did not wait to line up, but ran down the three flights of steps, grabbing on to sisters and brothers. Teachers lectured to students that they should control themselves. Many high spirited youngsters became sullen; quiet children turned hostile. Cooperative children were often marked as troublemakers and when they rebelled and felt discriminated against, they were forced to attend other schools and were tagged "fire victims."

We were all suffering from the shock of losing friends and having something taken away from us. The many facets of the disaster were so massive that we were frightened and our burdens were too heavy. We compared notes as to how much we suffered, as if pain were measurable. We witnessed parents struggle, unable to answer questions. We stood unprotesting as parents began separating and treating us differently until we became strangers at the dinner tables. We doubted we would ever again invest energies in friendships.

We were at the beginning of adolescence, but our puberty was exploited by the necessity of having to face mortality. We were numbed, we were shifted about, we were either ignored or smothered in affection, and we were different because of our experience. We were teenagers, wanting and needing acceptance, experiencing the perplexities of our sexuality, looking eagerly towards high school, and having to learn the sorrows of mourning.

HOME CAME A STRANGER

It was all sad because once we had been normal teenagers rushing pell mell towards life. Our experiences separated us from our peers; we were afraid to reach out and make friends, to search out our identities, to trust others, because we were afraid that suddenly everything would be taken away from us. We could not trust other people because our emotions were drained from us; we had suffered losses we should never have had to experience at the start of our lives.

Many of us kept silent, our condition being very unique. However, adults tried to handle us as though being scorched by fire was an everyday occurrence. Many children wondered why they lived and others died, but no one permitted them to talk about how they felt, as though silence was a protection against memories of what happened during the fire. We witnessed one of the worst disasters in the United States and were left benumbed and grieving. We were curious about the fire—we had seen not just people, but people whom we knew, burn to death. This disaster did not have the luxury of being impersonal like other disasters seen on T.V., or on the front pages of newspapers.

There were no herioc figures. We were simply young friends and teachers who had shared much of our lives together, and suddenly some were gone. When we rushed around in panic during that fire, we rushed to children like ourselves, to our friends. The last time I touched Helen I gave her my hanky to cover her mouth. There were many children who were helped on ladders by others. There were children holding each other, grasping onto friends as they hung out of the windows. There were children who pushed others to safety by knocking them out of windows. We stood comforting each other by saying it must only be a wastepaper basket fire, as we choked and coughed.

Nightmares and sleepless nights became common to our community. By day we were able to silence our fears by the expressed common opinion that the dead were saints and in the glory of heaven. At night we children were troubled by wondering why we were left behind. If the children who died in the fire had become angels, those of us left behind to bear our suffering must have been unworthy, unacceptable to God. Some of us tried to compensate for our guilt and tried to work this out in many individual ways, but much of our enthusiasm for life was dissipated in despair when we saw how adults all around us reacted to the situation. They never really understood what was happening to us.

We were different because of our condition and many of our needs were not met. One time the Red Cross sent all the children survival kits containing soap, wash cloths, and pencils. We were enraged they thought this would aid us. Our needs were bigger than the token gifts of the Red Cross emergency plan. We were shocked to hear adults rail against God; they complained about all the money spent on the dead and the injured; they talked about the dead children as though they had become spiritual identities and were not human anymore. We tried to repent for sins never committed.

We were instructed again and again not to cry, to be brave, and to hide our feelings. Some teachers at the various schools instructed their pupils to be careful of the tough kids from Our Lady of the Angels. Oh, we were developing quite a reputation all right; we were tough kids who were not supposed to cry!

It was a desperate time, and nearly everyone was going through some type of survival syndrome and needed help. It was an air of disaster madness that I went home to. I never walked down the street without being noticed or pointed out. I was an authentic fire victim. This opened another area of suffering I was forced into handling. At a time when all my energies were being sapped by the ordeal of merely sustaining life, I had to cope with the problems of mourning, rejection, disfigurement, and wondering why I lived.

CHAPTER XXI

It was decided, I was to go home. I did my time, four and a half months, with two days off for good behavior. I was leaving a most important part of my life behind me. There was an air of gaiety in the midst of all the confusion of hurrying to get my things together; then I had to wait for the doctor to sign me out. I was scared and nervously hesitant about leaving the seclusion of the hospital and going out into a world while I was still disfigured.

I tried to believe no matter how difficult the situation might be, I would find strength to handle it. I had learned the value of prayer, and even though I knew there were many troubles ahead of me, my faith would see me through. I kept repeating to myself that every person is only given problems he can handle, and not one more. There might be times I would get confused along the way, but I could always pray to God. "No matter how badly you are hurting, Michele, you can always go outside yourself and be kind to others," I told myself. I had learned my lesson well. Present pain was always the worst, and past pain must be resolved.

My sister, LaVelle, offered to come for me and take me home. When I was born I was her tenth year birthday gift, and fourteen years later, she was taking me home again from another hospital on her birthday. There was a short wait for a cab and several people gathered around to say good-bye while I was sitting in the lobby in a wheelchair. Soon a cab driver put his head in the door and called "McBride," and I was helped out of the chair. The driver took my suitcase and my sister helped me walk to the cab.

I took one look at the back seat and realized I could not climb into the cab; it was the front seat for me or nothing. I could maneuver the front seat, but first someone had to pick up my legs and squeeze them into the cab. I was worried because after all the months of training people how to move my body, I was once again trying to explain the best way to pick up my legs and what parts could not be touched.

I was on my way home, but I was so busy bracing myself against all the jarring movements of the cab, I didn't have a chance to watch what we were passing. Each time I winced it made me feel so inadequate, and I hated being dependent upon a strange cab driver.

The home I once felt so comfortable in had many obstacle courses for me to overcome. I was a stranger to this old familiar house and this made me sad, and I was resentful because I could not move about freely. Chairs, beds, stairs all became challenges. Rugs on the floor were especially difficult to walk on because I could not pick up my feet. I used a cane to brace myself from falling and it got stuck in the rugs; while tile floors were impossibly slippery.

People proved to be the biggest obstacles I had to confront. The affection and the encouragement of the hospital staff was soon replaced by the curiosity of ignorant people wanting to see what a "burn" looked like. They came to look at me, and what was so strange, they came for such varied reasons. Some came wanting me to console them and thus be comforted in some mysterious way. Some came hoping they would be relieved of their great burden of mourning their dead children. Others came out of curiousity, just to see what we children really looked like.

The reactions of those people to the sight of us also varied.

HOME CAME A STRANGER

Many gasped, other avoided eye contact, a few shyly tried to comfort us and others were unable to express their feelings.

I noticed superstition had begun to play a large part in their lives. A few parents made altars and prayed to their dead children, as though they had been transformed into saints. Some suppressed their feelings and found consolation in comforting the burned in many peculiar ways. These were the people who came and asked strange questions about their dead, as if I had a special link with them and the spiritual world and knew how safe these children were. I could tell them nothing, but found myself repeating the old cliche, "The dead were out of suffering," but where did that leave me and the others who still lived?

Whenever we burns shuffled along trying to walk down the street with our massive scars plainly in evidence, we were greeted in many different ways. People would run away screaming from us. Now I can understand the horror we must have been walking with maimed bodies, limping down sidewalks. We were the living, constant reminders for those around us who were trying to forget the fire ever happened. So they looked, then either ran screaming away from us, or began to sob uncontrollably. I was back in a community locked into and trapped by suspicion and grief. The changing of the seasons emphasized the fact that the children were missing, not only the children who died, but those who survived were also lost children.

It was a time there should have been great joy and excitement in my house because it was my eighth grade graduation. None of us could enter happily into the spirit of the occasion. We knew the graduation festivities would be marred by many illnesses and separations, that it could not be the joyful occasion it should have been. All my life I waited for those glorious streamers of powder blue and white with Our Lady of the Angels stamped in gold on them.

Suddenly graduation did not matter anymore, so I never returned to school to finish eighth grade. I wouldn't be able to walk, and I certainly could not climb three flights of stairs where the classes were housed. I refused to be carried. I tried not to be too disappointed.

The longing and the searching for my classmates, that began in the small dark cubicle at the hospital, continued after I came home. When I was in the hospital, I was shown a Mass card listing all the names of the dead children, so I knew the names of those who died. I promised I would not cry if I could see the list of the dead, and I did not cry. I still remember the shock of seeing the names of the friends I liked on that list. I pushed the reality of what dead meant out of my mind. I thought of my friends as being somewhere else — and that eventually we would be reunited, we would all be fine. I kept having the dream that now I was out of the hospital, now I would find the other children, and perhaps, my skin.

I was amazed at the cross section of children who perished; so many names seemed to be out of place on that list. The fire was impersonal in selecting its victims; there was no discrimination. There was one name especially startling on that list. The student had been transferred to our school the year before, and I thought, how sad not to even be a long standing member of the class and to have this happen. It was so unfair.

My longing and hoping to return to my class finally happened in time for the May coronation. There for the first time since the fire I joined my class, the students with whom I shared so many processions from the very first march in kindergarten to the eighth grade. I sat and watched them march down the middle of the aisle to the sound of "Tis the Month of Our Mother," a popular church hymn dedicated to Our Lady. The solemn students, marching and keeping a pew length apart, were erect as they filed into the pews. The girls wore white high heels and boys white shirts, all looked very smart. We were happy to don our caps and gowns, because we were finally the eighth grade graduation class of 1959, but the class was not complete; two of our members were still in the hospital.

As the class stood singing, I was escorted from the side of the altar to my place and it was the first time we saw each other. We did not turn or break rank. That would not have been allowed, but we wanted to hug each other. We were silent as we sat together once again. We looked at each other from the corner of our eyes, keeping our heads perfectly straight. We wanted to reach out and touch each other to make sure we were actually there.

Our ranks were thinned down so much, and for the first time there were more boys than girls.

I sat next to the girl who normally was four persons ahead of me and I was now flanked by another girl who was not my usual partner. The names on the Mass card started to mean something very much to me, and to everyone in my class. We were children, yet we greeted each other like veterans of a battle and this set us apart. Although young, we were old and weary, too. Many of us were nursing wounds and most of us were guarding thoughts.

I sat in the back of the class and started counting familiar faces, but the girlfriends I liked most were not there. I realized my classmates and I were confronted with the same sad problem—we would have to find new friends.

It was a time of mixed sorrow and happiness, but then, so much of life is.

The pressure of my peers made me practice walking and climbing stairs. I wanted to rid myself of the cane so I could walk unaided down the aisle of the church for my sister's wedding. I practiced walking the length of the house without the cane in order to be able to handle the long aisle of Our Lady of the Angels Church carrying a bouquet. The day of my sister's wedding, I walked down that aisle and I do not think any bridesmaid ever worked harder than I did to have that honor. There were blood stains on my bridesmaid's gown when I reached the altar, but I did not care. At least I could walk.

I was too young to appreciate my accomplishment that day. All I knew was how hard I struggled to get to the point where I could do it.

HOME CAME A STRANGER

Jubilant over my triumph in walking down the church aisle, I began to believe I could go through my class graduation as well. I continued practicing with increased determination.

Graduation came and it was anticlimactic for the class. We were torn apart and many classmates were missing. We had buried one fourth of our class. The members of our depleted graduation class sat in Our Lady of the Angels Church that bright June day with the sun shining so boldly on the stained glass window making it look like a greeting card design. We felt the warm breeze blow through the church as the hum of the upright electric fans circulated the summer air. The church was packed with many guests who settled in their seats, and the snapping of ladies' handbags could be heard as they reached for cotton hankies to wipe their brows.

It would be the last time for us to be together as a group. We were saddened to be the special class of students who had shared one of the most dreadful disasters in the world. Many of us had been together since kindergarten where we shared the happiness of having a live puppy in the classroom. We had learned our prayers and studied our catechism together. We learned the value of praying. We had tested our faith together in the fire, and our experiences would lead us into different growth patterns and would shape our lives.

We went through the normal rituals of graduation—autograph books, ribbons, class pictures and even standing in awe of the teachers who demanded that we march up the aisle in perfect formation even though some of us were limping. We would all think of those who were not there with us and we would feel this loss. Because of selective circumstances we would be honored as witnesses of such courage, fortitude, hope and endurance as was only written about in books — we would be witnesses in first person. We were the class of 1959. We were proud to be there together and felt stronger for having crossed each other's paths. We had seen each other in the worst of times, and we saw each other in the best of times. Often it was at the same time.

HOME CAME A STRANGER

I was presented with the highest honor of the class, the American Legion Medal. I was not happy at this honor, because it only pointed out once again that I was a fire victim. If the fire had not happened, the medal would have gone to someone else. I was sure those awards were given to certain types of students. I was not that type of child, unless that year the honor was given to the person who pinned the most bulletin boards. Then, I would have been a natural choice. I felt I was given something that belonged to the dead members of my class. It was also very difficult for me to get up off the seat and struggle forward to receive the award.

Much of my recovery was dependent upon the aisle at Our Lady of the Angels Church. It was probably good it was as long as it was, because it gave me the distance measurement of determining how far I could walk during the May coronation, my sister's wedding, and my graduation. "God writes straight with crooked lines," and I truly learned the full meaning of this proverb that summer.

The summer brought another sorrow, our last hospitalized classmate died. It was Gary, with whom I shared the hospital birthday cake. After those long horrible months of suffering he died, and the statistics of that cold December day's fire rose to ninety-five. His death triggered the opening of old wounds. I found out about his death the day I called his home to inquire about him and spoke to a family member who said, "Did you have the T.V. on last night? He died."

It was amazing how in his short life, Gary had touched and inspired so many people by the love he had for his fellow men. His repeated thoughtfulness enriched all of us who were fortunate to have known him.

It was difficult to say all those trite, glib things, that he would be safe and forever out of pain when we had all grown to love him so much. Now it was simply the end. There was no justice left, and I began to wonder if anyone died of old age anymore. His memory should be as brilliant as his short life had been, but there was nothing adequate to say. Even praying became impossible; to ask for acceptance of his death would have been fraudulent. I wanted to know why I was angry at God when I prayed and I wanted to know why anyone had to suffer such agony.

I wanted to keep something of him alive, because we shared so much of the same pain, but again I was too bitter to mourn. Now I had no one to look up to as an example of how much the human spirit could tolerate. The limits were now erased, and I began to learn that there are no limits to suffering. I was resentful that a burn victim should have to linger on and on and *then* die. It did not make any sense to me.

Suffering is not measured by degrees or length of time, only by adjustment. It was difficult to say he was out of pain and safe, because by this time, I realized pain does not kill. Was death, I wondered, the only perfect form of not being in pain? Was that the only way to eliminate suffering? These questions danced in my head.

I was confused, but I thought it was brave not to cry when at this time crying was so very important. I could not cry because crying would not change anything. The same was true of screaming. Screaming did not stop the pain, and by avoiding screaming, I learned to bottle up other emotions. I was weary of being an example. I just wanted someone to give me permission to cry, but no one ever did. I was not alone in having these restrained, natural feelings, and acting as though nothing out of the ordinary happened. A parent who lost a child stopped me on the street one day and asked if I hurt. I did not really know what to say, so I just nodded yes. She stood looking at my scars, and then said at last she was certain her child wasn't suffering. It was not very comforting to me.

Many years after that summer I met a woman on a bus who asked if I was Michele McBride. I said "Yes," and she started to scream, "Why did you live and my son die?"

If she only knew how many times I asked myself the same question, making me very insecure about my own existence.

Part Six

The Lost Angel

CHAPTER XXII

Many times I told my family and friends that because of the fire I had been reborn, given a second opportunity at life; I never said out loud to them how frightening this was to me. I had grown afraid of living, but I had to learn to accept life again.

Severe burns are a traumatic, deteriorating affliction with tremendous physical and psychological implications. Burns are also the result of accidents which many times involve others close to the burn patient getting hurt or dying. Burns experience all kinds of losses at the same time, and there is an important need to accept these losses, to accept the loss of parts of your body and even to mourn them. I had to learn to mourn before recovery could begin. I had to accept life again. I was given a second opportunity at life and at the time, this frightened me intensely because I was afraid of living. I had all of the problems derived from these causes and in addition, more devastating difficulties. These troubles concerned not only myself, but other students who suffered the same bodily destruction I did. I had to learn how to mourn for my friends and at the same time grieve for my own afflictions, and I had to master the greatest lesson of all, how to accept such sorrow. There was no teacher to guide me in these lessons.

I took these responsibilities too seriously until my life was stifled by the burden of caring. At the same time, I was still experiencing a great deal of negative feedback because of my appearance.

Walking down the street for burns became a painful experience. We were met with suspicion, and revulsion, and awe all at the same time. It was almost funny, the mixed up stories that were circulated. There were many unkind statements made about us, and so many stories about me. Some people thought I had been brain damaged; some thought I was crippled; some thought I was blind. It seemed as if I were not hurt enough. People wanted it to be more.

Many people had the misconception that the reason I was hospitalized so long was that they were playing favorites because my sister was in the convent. People who did not have immediate children involved felt left out of some of the pageantry and constantly kept talking about the profit of the fire. There were a few who thought the children who were hospitalized received money and gifts for being in the hospital. I remember once a woman screamed at me that I got paid for my suffering, which really made me feel miserable.

It was a very difficult time and a most impossible home coming. I had to calculate every move I was making and be very careful to avoid steps and curbs because I could not yet walk properly. I was determined to walk to the stores as I did before. This was difficult since I still could not bend at the knee. I used to plan my route carefully so I could go where there were low curbs and fences by which I could support myself and not fall.

I had to take longer routes than before, but I was determined I would walk alone no matter what. I walked down the middle of the block until I found a post on the curb and then I crossed the street. On the other side I had to repeat my efforts and walk in the street to find another low curb, but I found it. At the alley I grabbed onto a concrete fence post. Then I stepped down from the high curb and crossed to a high curb on the other side where I pulled myself up by reaching out to the corner of a garage. This was the only way I could get to the stores west of my house. I had to walk down the middle of the alley to get to the stores on the east side since all of the curbs were newly installed and too high for me to manuever.

During all this calculating of my movements, I could feel eyes watching me from behind the drawn shades and drapes in the neighboring homes. Neighbors were curious about me and took great interest in all my strange movements. I do not think that during this period of my life there was anyone who could actually share the experience I was going through, because I had lost most of my friends and I felt their loss very much.

The struggle of walking for burns was not just locating low curbs and fence posts, but was made more unpleasant by people running up to grab at us or to kneel before us right on the street. We were unable to fend them off and were made uncomfortable by these suffering people mourning their dead through us. One time an old woman sat down on the street when she saw me and started to scream and point at me. I could not get out of her sight because I was losing my balance, and I frantically looked around for something to hold on to. A crowd started to form watching her sitting on the curb crying, and me standing there fighting rage, and trying to brace myself to keep from falling. I had to stumble and push my way through the crowd to get near a building so I could lean on it since no one came to my assistance.

Another time I was in a store with my girlfriend, and we stood paralyzed in horror as a man grabbed the scarf off my head to display my facial scars to another person. The hurt I was experiencing from these onslaughts was worse than all the pain I endured in the hospital when all the bandages were pulled off my body. Nothing was ever as painful as a stranger tearing at my clothes. I was naked in the hospital, but I was never ashamed; I was always treated with great dignity. However, this hysterical man made me feel embarrassed, made me want to run away and cry, but I could do neither.

My friend was upset too because neither one of us knew what

to do. I stood silently with my eyes closed and meditated on leaving my body, just like I did when all the pain in the hospital got too bad, I pretended I was outside the horrible situation. Now I was away from the protective shield of the hospital isolation, and the situation was very different and more frightening. I never knew when I was to be accosted by people reacting violently to my scars, and this kept me in a constant state of apprehension.

Whenever this happened, all I could do was struggle to get home. I felt very guilty because I was weak and not brave — and I was miserable. The conviction that I must be a bad person to warrant all of this suffering persisted. I kept telling myself I could handle all the pain, and then I prayerfully offered my pain for the benefit of the dead whom I believed were not resting in peace, but there was no let up in my anguish.

I tried to find consolation in sitting and wiggling my fingers and thinking that at least I had the pleasure of my hands. I could do things with them and that was good. Whenever I rubbed my hands together I could see the thick purple keloid scars covering them and think how ugly, but if I kept my eyes closed, I knew I was lucky to have regained their use. This satisfaction was counter-balanced by a feeling of guilt that I had never lost the use of my limbs as do some other burns.

I was apprehensive about many things and so many thoughts churned through my mind, I often wanted to escape somewhere. People said, "Oh, you're so lucky to be alive," and I interpreted this to mean that maybe the dead were unlucky. This bothered me; I did not want to think of the dead as being anywhere but in the very positive state of good fortune. I wanted to cry about the dead, but it seemed so foolish; they were safe and out of pain. I kept saying over and over they were out of pain, and they were not suffering. All of this reasoning became twisted in my head because I was still suffering and it didn't make any sense. Then I would start thinking the dead were the lucky ones because they did not have to decide what to do with their lives. Life became too big a responsibility for me, and I was not relaxed anymore. I was tired from all the anxiety and my weakened physical condition, but I kept striving to get well.

I had improved somewhat but there were still things I could not do, and this embarrassed me. I wanted to be like my friends. I did not want special attention but I often needed assistance in doing the little ordinary things of life. I still needed help to go up and down stairs and this made me feel ancient and out of place. No matter how much I practiced, getting out of a chair without armrests was impossible. I was sick with worry when I went to visit someone, and I thought, what would I do if they did not have the right kind of chair and I could not lift myself in and out? When I was alone at home I practiced and practiced lifting myself from a chair without using the support of my arms. I sat on my little brother's play school desk, which was about three feet high, and tried to pull my body up without using my arms.

It took me all one summer to learn how to get up from a kitchen chair, without the aid of pulling myself from the table. One knee could not bend and this made it difficult for me to go to public places like shows, and even to church on Sunday. My knees did not bend so my legs had to stretch out in the aisle. Sometimes I sat down only to find I could not move from that position, and I had to wait for someone to help me.

Going to church caused a commotion because I could not sit in the pews or kneel and people stared at me. I found I could not muster the energy it took, so I was excused from Sunday Mass and went during the week. All through high school, burned victims going to Mass on Sundays was a big event for the congregation. People drew aside as though we were going to explode, or sat behind us and whispered "There's a fire victim." This angered me because I was determined I was Michele McBride, a person, and not a "fire victim."

My friends were all dead and I was lonely. I was left alone and it was necessary for me to seek new friends. I have always been considered a friendly girl, but at that time I was terrified to call anyone friend, because I was so afraid of losing that person. It would take me over fifteen years before I could call my friends "friends." I always referred to them as acquaintances. I became accustomed to keeping people at a distance. There was once a period in my life when I would exchange my name only if it was absolutely necessary.

I was growing removed from life and afraid of letting my feelings show. I was guilt ridden about living. I was always being reminded of the many bad experiences, or stumbling into other bad experiences. I started to dislike myself a great deal, and I became very rigid with myself. It seemed the harder I worked at making my body limber, the tighter my mind grew with so much inward self-anger.

I thought about how pleasant death really must be. After all, I knew many people who died, there must be something very right about it. Dying became more pleasant than living and much of the time I was haunted by the sensation of the dead being close to me. This was not only when I was asleep; I was terrified by their presence when I was awake. I wanted to be free of this bondage to the dead, but I thought it would be horrible to forget them. It would be very disrespectful, and my guilt kept growing.

There was always the uneasy feeling that I was searching and looking for something that wasn't there. I was convinced if I expressed positive feelings, such as love for anyone or happiness, sooner or later they would be taken away from me.

CHAPTER XXIII

In the fall, I started high school. Eight classmates from Our Lady of the Angels attended the same Catholic high school. On the first day of school, the pastor came to an assembly in the auditorium and welcomed us to the high school. We were surprised when he requested all the students from Our Lady of the Angels to stand. We did not like being pointed out, but we stood and heard whispers behind us, "Fire victims." Once again we had to mask feelings and refrain from protesting and expressing our annoyance.

My scars were vivid, hard, and thick. I still had open sores on my legs, and I walked with a noticeable limp. My surviving classmates and I were facing the terrible ordeal of trying to be accepted as normal, and we resented being pointed out as a special group. We were to discover there would be many times people would encounter us and try to satisfy their curiousity by knowing "fire victims." We would never be able to explain the terror of that day, no matter how many times we were asked, and too, there were so many opinions of what actually happened.

People could not grasp how fast fire travels, nor how hot smoke could kill instantly. They could never understand what fighting for life is all about. We attracted as much public news attention and sensationalism as though we had done something glamorous. Those articles were difficult for some people to read. People still suffering from shock read the papers and were forced to think of their own mortality. They brooded over how sudden and unexpected death could be. They became aware of the many aspects of death and they wanted information. So they turned to us, the children who miraculously survived.

I had never read about the terrible after-effects of a disaster. After-effects are as devastating as the original catastrophe, because everyone involved, both by-stander and victim, is thrown into a crisis state of mind and must manage to find a way to go on living. Disasters were remote to me before this fire! I was untouched by Hollywood movies showing an emerging hero who rescued the heroine in the nick of time. I did not see too many heroes in our neighborhood, instead, I saw grief stricken people. I had to learn people are sometimes frightened and weak, and this was not necessarily bad. I had to learn much about life in such a compressed space of time.

In spite of all the hardship I was facing, I was growing from all my experiences. All the negative emotions I was experiencing were not entirely bad for me, but they certainly took up most of my energy, and it was a time when my strength was exhausted by just trying to get out of a chair. I was exhausted most of the time, and struggled to keep from crying when I did not know why I wanted to cry. I was calmly hysterical as I went about living again.

It was a bizarre holocaust, and the intensity of its effect was not weakened with the passage of time; somehow it was gaining momentum. People wanted to share in the drama, and to feel part of something unique even though they actually could never understand. They wanted to impress upon us how close to the situation they were and how near to the dead children, who were specially chosen to merit heaven's blessing. Statistics about the fire were quoted to us. People forgot that our friends made up those statistics. All the names listed in the newspapers had meaning for us; they were our friends.

THE LOST ANGEL

Always I was to be reminded of December 1, 1958. Strangers often asked what parish I was from. Living in Chicago, the largest Catholic Archdiocese in the world, people usually ask what parish you are from soon after they obtain your name. It's to identify your neighborhood.

I always hesitated to say Our Lady of the Angels because that would start a long harangue on what they were doing the day of the fire and how close they came to the danger. Frequently, I was asked why I did not jump, and people were surprised when I informed them, "I did jump."

Those children who managed to escape uninjured were thought to be a novelty because the fire was so extensive, and they were quizzed on how they got out. There were no explanations for many things that happened, and there were no answers to many questions. We did not like it when people told us how we should have reacted, nor did we enjoy hearing what they would have done. It was impossible to describe the fury of a fire that showed no priorities, followed no set manner in its travels, and to which there certainly were no prescribed reactions to its presence.

We resented strangers telling us what they thought happened, and it made us sad to be reminded of December 1, 1958. We were confused, and our hurting was compounded by people trying to learn gruesome details they would never have the capacity to encompass. The fire may have been as exciting to them as an adventure story, but to us it was part of our past, and a very personal part it was.

I think all of us, both injured and non-injured, during high school had outbreaks of anger whether it was over the constant fire drills we knew were useless and outdated, or whether it was having to read stories about the fire. There were always glowing stories about the fire and what a wonderful job was being done to fireproof schools, but it wasn't true, or at least did not matter to us any more.

High school for me was difficult; I kept plodding along, trying to improve. I had trouble with stairs but I was determined to master them. I still, after all these years, dread stairs, but I do not avoid them. My appearance continued to be a source of much of my difficulty. My scars were so raw and sore looking they made people uncomfortable to be around me. The keloid were so much in evidence, some students would not sit near me. This was painful, however, I was still able to cope with this discrimination, not exactly with ease, but by successfully developing other abilities to overcome my physical handicaps. Once again, I did it by being friendly and going outside myself and with laughter. I was blessed with a fast wit and the ability to make people laugh, but I had an awful time accepting people liking me since I always thought they pitied me.

It was embarrassing for me and my friends when a stranger stopped me on the street to ask what happened to my face. This startled all of us, and everyone felt bad — and I never knew exactly what to say. People were crude and said things "Are you contagious?" and "Should you be out in public?" Once someone told me I should stay at home and not frighten people on the streets. I tried to force myself into believing these people did not mean to be cruel; they thought they were giving me good advice and watching out for me, but I really never succeeded in convincing myself of this. Self doubts kept growing and my physical presence was a constant reminder of a tragic day.

During this time there were improvements in my condition but they were tediously slow in coming, and I was young and impatient and wanted everything to be perfect. I wanted to be able to improve overnight. I wanted to walk down the street and not be noticed. I wanted to stop feeling the presence of the dead.

All the while I was fighting so hard for acceptance, I was learning to adjust to many handicaps I had not anticipated. My body was tired all the time; it took all of my will power to make it through a day. At night I did not find rest; I was troubled by a longing to find the dead.

Sometimes I felt miserable, and I was always on the verge of tears. I was weak and I had to make adjustments to many activities. Even sitting in class for six hours a day was fatiguing, and I had to worry about my legs getting bumped. It was both a physically and emotionally draining time for me.

I wearied of asking for special permission because I did not like emphasizing how different I was. I watched the effortless racing around my friends did on their tireless, strong legs, while I had to calculate the best way to shuffle along and then stop frequently to rest my legs. I remembered how easy it was to walk before the fire. Sometimes at night I would rock my body in despair just thinking of what it used to be and how free I was. I appreciated walking, no matter how difficult, and I tried to comfort myself by recalling how miserable I was when I could not walk at all. Then my spirits lifted and I knew it was good to be able to move, no matter what effort it took. I was still able to walk and I triumphantly could wiggle my fingers.

The smallest activity my aching body could perform meant so much to me, but I never dared tell anyone what an accomplishment it was. They would not understand, and so the people around me rushed by and never thought how great it was to be able to walk. They would never have the wonderful experience of thanking God for walking, a natural activity for most people, but a gift from God for me.

There were times the goodness I felt about my recovery was pushed away from me by misunderstanding people. Some people were not able to accept my scarring and reacted so violently I began to feel desperate about living. There is much personal satisfaction in overcoming a burn, but this was constantly being taken away from me. I am basically a happy person and I wanted to be joyful. Now I was growing terrified and I wanted to run and hide. I kept thinking the new problems I had to face were a punishment; I must be very bad to have all this happen. I began connecting the reality of my problems with a supernatural, angry force. The growing belief that the fire claimed for death only the good and deserving did much to destroy my self esteem. I was alive, therefore, I was bad.

People were screaming at me and I was screaming inside. There were enemies all around me, and the worst enemy was the one growing inside me. I was no longer loving myself, and I was turning out to be my own worst enemy. I do not know how I managed each day, but I had been instilled with a great sense of responsibility and I strove to go on just a little longer. Many times I succeeded in going outside the horror and I was able to relax and be almost normal. On all sides I was being told how to react, and I became separated from my own feelings, and this was dangerous.

One time a hysterical nun, who must have been suffering great guilt, cornered me and told me I should go see a priest about what happened to me during the fire. Everybody in my room was helped out, she said.

Was my mind failing me? Why was all this happening to me at a time when I was most vulnerable? I was being pulled apart and pushed and shoved. All the adults were treating me as though nothing abnormal happened, while I was living through these grueling experiences. I often felt if I could only die and stop this hatefulness everything would be fine. Then I went into spells of self doubt. I started to think I was an evil person who had to endure many different types of pain, that I had to work off some great debt to find eternal peace. I had witnessed hell. I had witnessed dying. I was still living but in a different kind of hell.

I went to bed nights and tried to find rest, but often I heard voices calling. This frightened me so that my nights were disturbed and my days were so long. I was not resting either awake or asleep. I started distrusting people and myself. I began to believe we must all suffer great hardships to prepare ourselves to be worthy of death. Then I became terrified that all those children had died too young, and were not resting in peace. I remembered all the horror of my wishing in the hospital that I were dead to stop the pain; now I was wishing the same to stop another type of pain. I could not believe I had endured all the suffering of recovery just to be faced with such horrible feelings, and I did not know what to do with them. I felt very guilty about the whole situation.

I often thought I must have descended into hell and was returned to earth after seeing many screaming, burning people. I could feel the bits of my stomach wrap around me, and I had a great urge to throw myself out of a window to stop the sensation I was falling. I wanted to stop falling and I could not. I stood by windows and heard voices say, "Jump, you did it before, feel yourself fall. You will stop the next time and you will never have to fall again. Don't be afraid to jump, the fall will put an end to the pain."

I developed a nervous habit of screaming that startled everyone around me. I did this whenever I was thinking all those morbid thoughts and someone walked up behind me. When anyone approached me unexpectedly, I screamed at the top of my lungs and nearly scared everybody to death. I trembled with anxiety when I started getting different kinds of psychic messages which made me mistrust my own mind. Because of all the panic I lived in, I was convinced I was doomed for hell. Certainly hell was no worse than what I was going through.

People continued to remind me I was lucky to be alive. This statement always irritated me because life is not like winning the "Irish Sweepstakes." Everybody who is alive should be termed as "lucky," not only the maimed.

CHAPTER XXIV

The burden of pain coupled with my fear of approaching madness finally got to me and I had a breakdown during the end of my senior year of high school. I was emotionally and physically exhausted and collapsed one day in school. I spent the next three weeks in the hospital, but nothing was done to relieve my condition. I was often sullen, and wore myself to a frazzle making all types of cries for help, but no one knew my trouble stemmed from my suffering extreme guilt and depression.

A psychiatrist came to see me one time, but instead of talking to me, he took my girlfriend outside in the hallway and asked her what happened in school to make me collapse. That is the nearest I got to psychological attention during those three weeks.

I started having violent nightmares, and it took several people to keep me in bed while I threw myself around. I was very frightened that these spells would continue. However everyone behaved as though this was just an adolescent stage I was going through. I prayed that my life would end, and I was really very angry when my prayers went unanswered. I thought I was being held prisoner in my body and that everyone was interferring with my dying; I felt there was a conspiracy to keep me from being at peace and safe with the dead. Also I was perplexed with the problem of what to do with my life because I was convinced I had been given a second chance at being born, and that doubled my responsibilities toward life. I had to restructure my whole life from beginning to end, and between adjusting to all my physical complications, I was examining my existence. I realized I was becoming too serious about life and my continuous searching for my identity made living a great hardship. I reached the limits of my endurance and could not relax. In spite of all that was churning through my troubled head, I presented a very pleasant outward appearance most of the time. This was confusing to my family and friends. I had always been high spirited and playful, even while I was in the hospital; now I could no longer deal with all the inner anxiety I was going through. When people got too close to me, I grew fearful. Someday they might find out what a bad person I was, and I kept wondering why it was people did not notice how truly bad I was. The fears I locked in my mind were far more damaging to my psyche than any of the scarring of my body.

Why I wanted to suffer I do not completely understand, but I felt extremely guilty about living.

Surviving a serious burn was one problem, but having been in Our Lady of the Angels fire was another. After high school, looking for a job was difficult; not only did I lack experience in office work, my appearance did not match the decor in some offices. I found out what job discrimination was. There were times people told me frankly I could never have a front office job because of my looks. Somehow, no matter how many times I was rebuffed, I was always able to go outside myself, even during the gravest day, and I was able to find something within myself on which to cling. I learned to treasure the humdrum parts of life as my accomplishments.

Most of the time I stood alone during those days but I still found joy in the world. I strove always to go outside my internal woe and to be generous and helpful when needed. All the terror I was experiencing in recovery did not make me stop being genuinely my own person. I did not call anyone friend, although there were those who called me their friend. While I thought I had to suffer to obtain eternal peace, I continued to be supportive of others. I could not understand all pain, but because I did suffer, I felt how others felt.

I was very unorthodox and unconventional in my sense of humor, I was still popular and people did like me, but all this did was to make me uncomfortable. The combination of terror and laughter made me inwardly explosive all the time. For a while I was able to conceal this feeling and present an outward picture of a person who was "all together," as the saying goes. Had I only fathomed that my real problem was my need to give vent to my mourning—I was literally drowning in unshed tears—perhaps much of my inner stress would have been relieved. I needed the healing balm of normal mourning, but I was unaware of this feeling — and anger continued to rage within my soul. I became especially confused every time they celebrated the anniversary Mass of the fire. It was a time sermons were inappropriate; priests, without real insight, spoke from the pulpits and admonished us to be brave and not to cry because the dead children were assuredly safe in heaven.

I often wondered why parishioners were being instructed to feel nothing, instead of letting out their sorrows. Pent up emotions destroyed some families. Years after the fire and even to the present day, there are some terrible unresolved feelings. There are still some parents who walk across the street to avoid survivors who are now adults.

I lapsed into a state of shock that continued for many years, and I do not know how I carried it off. I kept searching and thinking out problems and I had only myself to rely upon. My feelings of guilt persisted and became monumental until I grew weak and not strong enough to bear them. Childlike, I believed all the stories about saints and martyrs and knew that as much as I suffered, it was not enough to compensate for all my imagined evil deeds. Foolishly, I never thought to ask myself exactly what evil I had committed. Suffering became a way of life for me, and I was afraid to change. I guess I thought any change would mean I would have to stop laughing. As much as I was discriminated against, I'd rather be in my own maimed body than in a normal body whose owner had a dull perspective on life.

Every day seemed to bring a new crisis, until disaster became a part of life I did not want. I yearned that one disaster should cancel all other pain in my life, but that's just not the way it is. Damn! I thought, the children who did die were lucky, and I certainly am unlucky to have these problems. Thoughts of suicide entered my head.

I decided that if I could not resolve the questions that were bothering me, I would simply kill myself and find out what was on the other side of life. I became preoccupied with death. It never frightened me after the fire. I thought I was indifferent to death, when in reality, I had no feelings toward life. When I was critically ill in the hospital, no one would talk about death. I learned this was because people were so frightened of dying they found it impossible to come to terms with death.

When I felt threatened by friendship and commitments, I withdrew and detached myself from life. I could not imagine anyone liking me when I was so ugly and a constant reminder to them of that tragic day of the fire. I must be very evil to have become a symbol of the tragic event.

I was also giving out double meaning statements that bewildered everyone. Once on a date the boy I was with asked me what I was planning to do with my life. I said, "Kill myself." There we were in a crowded bar and he pushed me very hard, and yelled at me never to say that again. It was the first time someone ever heard me call for help, and I was stunned by his reaction. I decided at that moment never to say that again . . . at least to him.

Violent nightmares continued to haunt me. I would wake up fearful because I could hear my own voice saying I would never be anything but bad. I would just have to join the dead and it would be so very easy then. I would not have to worry anymore, and then no one would know what a bad person I truly was. Those nights I woke up and rocked myself back and forth and thought about ending my life. I was afraid and I was tired of people who kept associating me with the fire and telling me I was brave to have survived. I worried about my limitations that would never leave me. I still had trouble walking and my knees were so weakened by the burns I had to be afraid of their collapsing on me. The sensitiveness of the keloid scars would always react to conditions of the environment, and I would always have to take this into consideration. Those were the problems I did not expect.

I was becoming disassociated in my feelings. People were saying I was brave, when I never had any choice in the matter; people were telling me I was lucky, when I was very weak; people were telling me that I was ugly, and people were telling me I was pretty, all at the same time. I stopped trusting my own feelings, and I was lost.

I wanted to please everyone, and have people ignore my scarring and accept the real me. If people did like me, I did not trust them; I always thought they were giving me pity. I wanted to stop time, and think things through. I wanted to stop for a moment and try to catch up on life that was rushing by me. I was dangling and I felt I could not hold on much longer. I was being pulled one way, and being delivered another. I needed to change, but I did not know what to change, since my choices were always limited by the fact I was permanently damaged. Nothing would change this. It was maddening.

I wanted to run, but I was afraid I could not run far enough. I dreaded the night, but welcomed the dark. Then, because of the turmoil always in my mind, I waited impatiently for the comforting light of dawn which never came. The unhappy thoughts were always with me. The only way I could stop them was to stop living and end my agony.

CHAPTER XXV

When I became of legal age, eighteen for a female in the State of Illinois, I filed suit against the Archdiocese of Chicago because the Fire Fund that was paying our medical bills was supposed to be closed out when I turned twenty-one. My parents did not have to worry about the cost of my medical care as long as these expenses were paid, but the news of the Fire Fund being closed made it necessary for us to file suit.

There were many insulting comments made about our suing the Catholic church; some people thought it was like suing God. It would have been easier to sue God, or at least the judgment would not have been tarnished by the human element. The legal suits were all to be handled at one time, consequently it was necessary to wait until the last injured or dead male child turned legal age before the suits were taken to court.

Like the investigation of the fire, the hearings were kept at a high emotional level, and they came during the time I was examining my existence and not very certain of life.

I always felt the legal aspects of the fire were treated like a scholarship fund by the lawyers, judges and court systems. Compensation was never explained to me and compensation to me always meant taking care of the medical bills that might arise.

I remember walking into a judge's chamber and being asked to be seated. I was reassured there was nothing to be frightened of, and I replied that I had faced more difficult circumstances than just sitting in that chair. One judge amazed me by asking what I thought of the "Our Lady of the Angel's" fire, and I said I thought that "the children who died were lucky. I have been given a second chance at life, and I feel this has doubled my responsibilities to life. If the judge will please tell me what I should do with my life, I shall be very happy, because I do not know what I should do with my life now."

When the settlements were to be made, the courts wanted to disburse the monies to the burn victims on one of the anniversary ceremonies of the fire. I would not appear and the date was changed. I could not stand hearing one more touching story about the fire. None of them had any meaning for me.

There were many basic conflicting personality traits thundering inside me, my life became a storm of gloom and doom. I could not forget the dead. I could forget the horror of the fire, but not the dead. I could feel their presence, or was constantly reminded of them. When people asked questions about my scars, I never thought of saying "none of your business." I thought I owed them an explanation, as though my body were not mine anymore. If I planned to be in a group of people, my reputation as a fire victim preceded me. Then people sought me out to relate what they had been doing the day of the fire, or that they attended someone's funeral. Inevitably, it was the last rites for a child I knew well.

What do you say after a child has been dead twenty years? I'm sorry I'd rather not talk about it? Don't they have the right to be remembered? Their life span was so brief, but they deserve to be remembered. I think that is why man fears death; he fears he will be forgotten. That is why people have children; at least someone will remember them.

We may be tyrants, we may be peacemakers, but we all want to be remembered, it is an inborn desire of the human race. How could we face each day with the threat that no one would ever know we existed? How totally futile our lives would become to think all of our qualities would vanish with us after we were dead. But what happens when a child dies?

The sadness of a child's death is intensified because of the absence of tell tale signs of aging, such as wrinkling, graying hair and loss of energy. Instead there is a bundle of energy exploding, waiting to test his own strengths, when the impetuous hand of fate snaps life away.

Children, who were here on earth momentarily and then snatched away, belong to the ages. They missed out on living. I could realize the sadness of their death, and that made mourning more difficult.

I think now there were many times I wanted to forget my dead friends, but I did not know how. I tried to forget only to become besieged with constant reminders.

I began thinking of myself as two people. Often when people called to see if I were home, I said, "We'll be home" and if they asked who was with me, I just answered, "Oh, nobody but body and soul." I firmly believed there was a good me, that was generous, and a bad me — and I always had to hide the bad me or people would find out what a terrible person I really was.

Scarring continued to be difficult to handle, and if I met someone at night I tried to avoid meeting them during the daytime. It was all right if first we met in the bright of day, but I did not like the difference of the two meetings. If I chanced to meet someone, and they did not ask what happened to my face or hands, I became worried. If people thought I was pretty, I wanted to shout, "I'm not, can't you see I'm scarred?"

I became filled with mistrust and so pessimistic I anticipated fatalities happening again and again. If I gave a party, when everyone went home I was happy the ceiling had not fallen. This combination of gaiety and sadness constantly churned inside me, but outwardly my appearance belied this confusion. I kept fighting these emotions, trying to push them aside; I could not resolve them. Time may heal wounds, but it certainly does not necessarily resolve bad feelings.

I entered a period when I was a friend to those around me, but a stranger to myself.

I did my best and got front office jobs. I worked in the field of advertising since I liked working with people. One of my jobs was as a booking agent for models. It was in this job I became interested in using makeup. The Chicago area models told me their beauty secrets of how to apply makeup.

I was an editorial assistant for *Billboard* magazine, and loved the job. I affectionately called my bosses the "odd couple." They were often called on the carpet for my high spiritedness. People in other offices could never figure out how we could have so much fun working, and get out twenty-two pages of a weekly magazine. Once again my laughing was offending people. There was a budget cutback and I was forced to look for a new position. I felt I lost something very important to me, and I was being punished for liking my career too much.

I moved away from home, had my own apartment, and dated like other women my age. However, I still had a haunting feeling that I had to do more with my life to feel worthy. I did volunteer work in both St. Anne's Hosptial and the Cook County Burn Unit. I had a secret desire to work forty thousand hours of volunteer work to pay off the hospital bill that had been paid for me. I never felt comfortable with all the money spent on me. I always forgot I had nothing to do with being in the hospital. In spite of my cheerful exterior attitude, I still had very crazy thoughts.

I did not know how I became so disoriented and sometimes I thought I was losing my mind. I kept changing my life style and finally I decided to go away from Chicago and enroll in a distant college.

After working for seven years, I became a college freshman at Southern Illinois University. However, all the craziness that surrounded me in the city came with me to college. I still did not know what was worrying me. At this time I was very frightened and on edge, and I was isolated from everyone and everything in the city. I was running but what I was running from was myself, and that I could not escape.

During my first year in college, I managed, but my anxiety steadily worsened and I was plagued by a dreadful feeling I was not doing what I really was supposed to do. I wanted this feeling to stop. No one ever really knew how much I was dwelling on death, and how often I contemplated suicide. I did silly bargaining with God, promising I would do so and so, if he would show me some sign that I was to end my life and escape all my misery. I still retained the ability to push myself and do things. I could function fairly well, but I was always on the verge of tears and I could not explain why.

I had hysterical outbursts and surprised even myself with my yelling and screaming at people, with friends or strangers; it did not matter anymore. The voices I kept hearing through the night and day kept me frightened. I was angry that the voices never left me alone and I really did not know where they were coming from. I cried and sobbed, and then my tears scared me. Brave people, I knew, did not cry. All I wanted to do was cry, and I was always frightened by the urge.

I was very well healed. The scarring on my face was softening and the color was good but still I dreaded meeting anyone because I had been hurt by so many thoughtless people. I needed to reenter the world, but I had been separated and confused for many years; I was in a quandry as to how to start.

I was withdrawing, closing out the world; it was frightening. My acquaintances noticed my increasing disinterest in life and desire to be alone. My one thought was to disappear and hide. I wanted to stop being.

One of my friends noticed this withdrawal and suggested I see a therapist. I said, "No, it would not change anything. I do not think that would solve my problems. Nothing could make me stop feeling this way."

I was hurting and I did not realize the pain I was causing

myself. I was ashamed of my feelings and I did not know what to do. Finally my friend talked me into making an appointment at the counseling and testing center at the university. I was calling for help which heretofore I always avoided. I was also screaming to be left alone; often it was at the same time, and this was terrifying. Mostly, I was determined I would never again allow myself to be happy.

Part Seven

Up From Depression

CHAPTER XXVI

I was twenty-six years old when I walked into the university counseling center, thirteen years after the Our Lady of the Angels fire. I could tell it was going to be a very hot July day because I could already feel the warmth of the sun's rays against the morning dew.

Now, when I think back to that summer day, I can almost feel again the grass brushing against my sandals as I walked toward my car to drive to the center. I was going largely out of curiousity and to please my friend who had urged me to go, but unknowingly I was ready to begin the long journey toward recovery.

I filled out a questionaire by checking the proper little squares: Yes, I was depressed; Yes, I was more depressed this year than last; Yes, the depression was growing worse instead of better; No, I did not know the cause of my trouble; the reason I was there was to find an answer.

That unanswered question was: Why was it that the fire had stayed with me all of those years and had become my master?

I did not realize how long it was going to take me to find the answers and rid myself of this persistent problem. I had the naive idea that one school term of eleven weeks would suffice for my problems to be magically resolved. I soon learned my problems were never going to be magically resolved, that I alone would have to work them out and find the answers, aided by counseling. There were days that were dark with depression when I felt listless instead of feeling well as I had expected. They told me this was entirely a normal reaction in cases such as mine, but this was scant comfort to me.

Because I did not trust or even like myself, I constantly dreaded facing the future, and became angry when I was told I would have to effect my own change. The truth was I did not want to face up to the changing.

Hysteria, anxiety, hurt, and guilt were the emotions that had filled my being during that long painful period of my life. I cannot recall having been calm and quiet for one moment.

One time I had made a clear decision and it was, that no one would ever have to pity me because of my scarred appearance. I had been wheeled to the sun parlor at St. Anne's Hospital, and a man came over and said, "You poor child." I thought, "No one will ever dare say that to me again." Despite my resolve to be strong enough to resist sympathy, I still became disturbed when anyone got close to me. I was convinced that anyone becoming precious to me, would inevitably be lost.

All through my adolescence and on into adulthood, I had problems facing any loss in my life, and I continued my childish belief and interpreted all loss as punishment for my sins.

There was the first time I drove a car by myself, and I was involved in a four car collision. I wasn't hurt in the wreckage, nor were others, but my escape was proof to me that I was being punished because I had displeased God in some way. Another time I was burglarized and a cherished charm bracelet was stolen. My father's baby ring was one of the charms, and the loss grieved me because he had given it to me. I wore this ring until it had to be cut off my finger and then it went on my bracelet as a charm. I wore this charm bracelet on my left wrist to cover the keloid scar, and when it was stolen, I interpreted this as punishment for my sin of vanity.

UP FROM DEPRESSION

I was always fond of wearing fine rings and wore my grandmother's diamond and cameo ring every Sunday. After my hands were so badly scarred, I stopped wearing rings because they drew too much attention to my hands and people kept asking me questions. I believe that of all the scarring I suffered, the scarring on my hands was the most painful, because they were always on display.

I believed that everything I enjoyed doing or liked would sooner or later be taken away from me.

My dear old, lovable, English sheep dog regarded everyone as a friend and followed me everywhere I went. One day, a distraught neighbor killed my dog, and I interpreted this as another punishment for my wickedness.

Now, when I think back on those days, I realize I must have wanted to suffer, and I certainly did it in a big way with no let up. All the energy I wasted tormenting myself during this time was so needless; I was exhausted and worn out. Now I wonder how I found the energy to recover.

I was injured in the days before the Beatles made Gurus popular, and positive thinking and mental therapy were beginning to be talked about. I had a serious mental health problem and there was no help available to me except to try to "hang in there." Hard as I tried, I kept losing ground. I was near exploding most of the time. All my anger was churning inside me, and I was about to destroy myself. I became my own worst enemy. I was much nicer to my friends than I was to myself. I could never accept my losses philosophically but drove myself into believing that everything in life was bad for me. The only escape would be for me to stop living.

I could no longer stand the daily torments of life and my problems kept mounting; so, I isolated myself out in a farm area. I wanted to make myself disappear. I wanted to be away from all the hurting of past miseries that haunted my thoughts and which I interpreted incorrectly. How quickly people were to say "You are so brave" until I felt the need to cry out that I was weak and timid and how wrong they were to think me strong and brave. I felt guilty that I was deceitful in not telling people the truth about myself. There wasn't one aspect I liked about myself, and I could not bear to have anyone like me.

I frequently denied I needed to change. In my defiance I raged because at the same moment, deep within, I realized the thought of changing was so frightening to me; my sole defense was for me to become defiant and hostile. I denied that life was worth living, and insisted everything was either bad or mean, and that there were no pleasures for me anywhere. Life was to be endured, not enjoyed, I insisted. Then, somehow, I found myself relaxing my guard and momentarily enjoying life. These brief moments of relaxation did not help but made life harder for me. I was so scared of saying "yes" to life, that I looked upon death as being exciting, or at least restful.

I carried the presence of four, dead, girlfriends with me into adulthood, and I thought their memory should never leave me. I often worried about them and wondered if they really were resting peacefully in death, until I sometimes burst out in hysterical screaming, shouting and shaking, because I thought I was obligated to do something to relieve their fate. I begged these specters to leave me alone and go away, but I could not stop them from following me.

In public I was always on the verge of tears and I cried privately, but weeping gave me no ease. The more I fought crying, the more I needed to cry. I remembered the sermons of my childhood telling us all not to cry, assuring us the dead children were safe. I can remember hearing "Don't cry, time will soften the pain before long." I had been told not to cry at a time when weeping would have been so natural.

People kept telling me I had survived the worst experience I would ever have to endure in my life, but I always wanted to scream, "No, it wasn't the worst." The most unbearable was the daily threat of people hurting me and demanding an explanation about my physical looks, and the need to mourn.

I did not trust anyone. I thought only perfect people were acceptable to society and since I could never be perfect again, I was fated always to be on the outside of life. If anyone touched me once, I vowed they would never have a second chance, because I was so repulsive in body and soul. I often thought I was doomed to go to hell because I felt I had been in hell, and for some unknown reason was sent back. I had a very strong sense of being in a different level of life, uncommon to most people. Whether I actually believed I had died and returned from the dead, I don't know. I knew I was marked and excluded from most of the human race.

All of this I found difficult to pour out to my counselor, and then I grew angry when an immediate solution was not forthcoming from him.

Another part of my craziness was I could not spend money on myself without feeling guilty, but I could turn about and give away large sums of money to ward off the guilt. There were not many aspects of living I enjoyed.

That often painful summer at the university counseling center stretched on into years of therapy, and I fought every step of the way. I refused to believe I could change, because I did not understand what needed to be changed. I did not want to change either, and so found that one of the advantages of being preoccupied with planning suicide was that I didn't have to take time to assume any responsibility. I think that was the part of my craziness which I enjoyed. It was fun not to have to think of the future. I devised ways to trick people into liking me and when they did, I devised methods to prevent their getting close to me.

I was often so melancholy during this time, that I hated people who were compassionate enough to see through my act — and I wanted to annihilate them. It was then I would do a quick change. I would work very hard on friendships believing that sooner or later the people would die, so I had better be good to them, or I might never get another chance. Also, I yearned to have people miss me after I died. This became an obsession with me, because it seemed mine would be such a valueless life if I were not remembered. I think in many ways I was living my life and always writing my own epitaph.

I was also under the belief, prevalent at the time, that therapy was a sign of weakness, and I would not admit to anything but being very strong and brave. I resisted therapy and was completely turned off by it. I fought with my therapist insisting I was well. That was when I knew for sure I must change my attitude, and that considerable pain would be involved in making this change.

I have suffered both physical and mental pain and I would not opt for either, but I always reminded myself the present pain was the worst, because that was the pain I was dealing with at that particular moment.

While I thought I was a terrible person for having to witness such terrible experiences in life, I forgot I also was privileged to see beautiful things. I refused to allow beauty to come into my life. I only wanted to dwell on the terrible. It was easier to be terrible than beautiful. My life was complicated because somehow many of my good qualities must have shone through.

I had a large following of people who actually liked and supported me, or the parts of me I would let them share. I must have scared the hell out of them when I sank into a black mood of despair and announced that I was going to kill myself, and then did a chameleon act and became cheerful and behaved for a few days as though I had not a care in the world.

CHAPTER XXVII

How sitting in an office for fifty minutes a week, often twice a week, could make me want to live or accept life again, was beyond me. I refused to accept the recovery because I was hurting and remembering the hurts of long ago. It was very hard to put Our Lady of the Angels fire into proper perspective. It was a very complicated time.

The post burn patient is often left to cope on his own. This is difficult because fires can destroy not only bodies, but many times there are multi-injuries involving others. There is also the loss of homes, possessions, having everything taken away without warning. In brief minutes you can be confronted with a crisis that most people will not encounter in a life time. Burns are like a serious illness, they cut people from you, and this is very difficult for the post burn to accept, especially the teenage post burn.

Burns are one of the few afflictions that attack the body from the outside, and the result is so apparent it frightens people so they do not know how to react. We "burns" must devise methods of coping by ourselves.

All through therapy I forgot how excited I was in the hospital over the possibility of my being able to grow new skin. Just as once I did not believe I could grow new skin, I also believed that I could never work out and rid myself of all the fears, anxieties, and guilt I was feeling.

It was a difficult time, but it was also a growing time. I was beginning to learn how to dissect the many aspects of death, to grow out of desiring death, and at the same time, to appreciate death and dying. I learned that what makes dying so difficult for many people is the fear of the unknown.

My guilt over Helen's death was not easily overcome. I continued to suffer great anxiety because I had never asked to be alone with my dearest friend when she was dying in the hospital. So many times I had wanted desperately to be with her, not to say anything, but to let her know I cared. I was afraid no one would have understood. How forlorn I felt because I had not demanded to see her, just to touch her and say good-bye. However, at that time, I was surrounded by people who were afraid of dying and they did not understand that dying was all right.

Helen and I had shared a very special boundary. When the harmonious structure of our universe was broken and the cosmos started crashing into eternity, we had undergone an extraordinary catastrophe of life. We had stood together for a brief twinkling on the edge of time and then were parted. Perhaps Helen was satisfied, but I had to learn how to re-enter the world. That awful experience enlightened me, but did not assist me in venturing back into life. I have gone through all the usual stages of dying so many times that I have learned our fears actually are not about the act of dying, but center upon death itself. The dying person may feel the loneliness of death and not be frightened until he senses the fears of those around him — an anguishing moment when the fear of losing a dear one is painfully evidenced. The institutional designs for dying in a hospital are so rigid that death is met as a negative force.

Through my incessant preoccupation with thoughts of death, I came to the conclusion that persons who are afraid of death have been afraid of living. Instead of death being thought of as a separate part of the life cycle, it should be considered a great event. I do not think death is morbid, only people make it so.

Death could be construed to be happy and fruitful, if only we were allowed to plan the way of dying. I mean more than going off to a lawyer's office and making a will. Such an act does not prepare us for death; it merely sets up a scheme to disburse our property.

To think about death makes us sad, makes us lonely, because we are scared of the unknown, but just talking about death does not hasten death. I shouted from the top of my lungs I wanted to die, but I did not die. I longed to die to escape great pain, but my will did not make it so. I courted death with suicidal thoughts, but I did not die. I studied death, I felt death, I screamed for death, and continued to live, but the important outcome from this unfulfilled death wish has been that now I am prepared to meet death on my own terms and without fears.

I sometimes wonder when I will die. Death and I have wrestled and fought many times, but so far neither Death nor I have been the victor, however, Death has left a mark on me. I am constantly reminded I am mortal; one day I will slip from this earth, and time will pass and it will pass without me. I only hope the earth is a little better by my having been here. I hope I have not taken too much from this world, and I would like to think I have given back a little something.

I hope people will mourn and cry and scream when I die. None of this balderdash about "It is better that she is gone," no matter how long I have lived. I want to be mourned and the more the merrier. I want a happy death. I hope no one dares say to my bereaved ones, "Be brave and don't cry" because they will need to cry, and I want them to be comforted by being free to weep and mourn.

I think saying, "I love you" is important in life, but to say, "I will miss you when you die" is also very important, however, we do not allow ourselves to reveal these innermost thoughts. I believe that people who are afraid of aging are afraid of dying, and this constant fear stops many people from enjoying life, because they are pretending they won't die. Pretending is no escape from the inevitable. Death should be included in the life cycle, not as the alternative to life, but the completion of a life when, perhaps, we journey into another world, taking all the knowledge from this life with us, and that is good in itself.

I wanted very much to be able to touch my girlfriends, just once, but I was a child. I did not know how separated all my tribulations made me. Now, I tell my friends that when they are dying, I will come and say "Good-bye." They all claim they will not want to know when they are dying. I tell them, "You will know." I try to tell them how lonesome dying is and that they will need someone, but they do not believe me, because they have not confronted death face to face. They do not understand how good it is to know that death is near and still be unafraid, to have someone hold them close and share the experience with them, perhaps even laugh and say "Good-bye," and let go.

Through the long years of pain and embarrassment from the day of Our Lady of the Angels holocaust, I had faced death and learned how to die unafraid. During those months of therapy, I was to learn how to live again and not be afraid—this was equally painful for me and I struggled every step of the way.

I had to understand that people are often frightened about many facets of life, and that normal people find it easier to deny problems than to face them. I was not alone in this. I learned in therapy that when people are thrown into catastrophic situations, they are often more fearful than courageous, and withdraw into themselves and stay that way. Many of the unpleasant situations I faced were not because I was bad and being punished, but because of frightened people who were too hysterical to help me.

This unwarranted belief damaged me to the point that I grew away from people and became afraid of them. When I was bursting with excitement at my good recovery, I had to go back into a neighborhood full of people still suffering from the effects of a great disaster. I was aware of their horror stricken faces when they saw me. I was a constant reminder to these people who had been admonished to forget and never speak of that horrible day. I saw families break up because no one could stand the suppressed emotional pressures. We avoided each other, and most of all we avoided the necessity of mourning.

Saints to children are far off people, who during the middle ages were stoned to death holding crosses with lilies growing on them. But these were our friends, our sisters, our brothers, and they all had names and human qualities and we did not want to forget them as being human beings. It was impossible to think of a favorite friend as a super-mystical being when we recalled going to the movies, or playing house, or baseball together. What really hurt the most was the fact we were not allowed to say how much we missed them.

Once, when I was eighteen, the mother of one of my dead girlfriends met me in church and asked how I was. I told her fine and that I was working. She told me she always noticed me in church and wondered if her daughter would have grown as tall as I. I told this mother I remembered her daughter always stood behind me in lines for communion, confirmations, and processions because she was a little taller than I, and perhaps, she would still be taller than I. It was such a relief to be able to speak of this girl, not as some heavenly body, but to remember her as human. It was so much easier to talk about her in worldly terms and it was not at all morbid, it made both of us feel good to say something normal and not pretend.

I hope when I die people will remember the style I developed in life. I hope they remember how much fun we had and the troubled times we shared, and I hope I am not completely forgotten. I hope I have not developed to be this crazy for nothing. I want to be remembered, to be mourned, to be missed.

CHAPTER XXVIII

How does therapy work? I do not know. I certainly was a talker, so why did talking fifty minutes a week make me want to start living again? How did I stop hating myself? I don't think I really ever understood the process, and it probably was more difficult for me because of my locked up, raging anger. I even hated the fact that I needed therapy. I was angry I had to do all this changing when I could list alphabetically, according to size, shape, color, other people who were much crazier than I, and they were not in therapy.

I must have been exceedingly unpleasant to work with. I was uncooperative, evasive, and angry, but I guess what really counted was I truly wanted to change, even though for a long time I would not admit this to myself. There continued to be dismal times when I could never imagine being happy. I did not know what it was to live without depression and guilt. I thought this was to be my only script in life, and it was a pretty miserable one at that. I was reluctant to admit that what I had to change were my troublesome emotional problems that caused all my unseen symptoms, which I felt acutely.

I was constantly upset it took me so long to change. I was discouraged and probably would have given up had not my counselor and friends encouraged me by believing I could overcome my fears and suicidal tendencies. I believe now everyone requiring psychological therapy must first realize he must work towards change, because of the constant threat that if he doesn't he will lose everything. Support from friends is good at first, but as the depressed person improves, support must come from the innermost part of his being. Growing stronger or closer to one's true self is often wearisome and disheartening.

I had to learn life's problems come not as punishment, but as a part of life to be faced. It is strange, but I never thought it was a punishment to be burned, but I did think for years it was a punishment to be an Our Lady of the Angels fire victim. Through therapy I came to accept that this world is made up of many forces, and at a time and place I met a greater force than myself, through no fault of my own. Fire overwhelmed and consumed my body and because of this, I had a complete breakdown of body organs. I came to believe this was not a punishment from God.

I do believe in God. I fight with God all the time. Some people might call this prayer, but I think that prayer is when you ask for something, and I rarely ask God for anything. On some occasions when fighting with God, I threaten to expose him to the world, or I question if God really knows how difficult life can be. I have often thanked God for special things in life, such as making water and coffee calorie free. There are nice light touches to this troubled world, such as sunrises and sunsets, but why is there so much work to be done on earth? I am basically a lazy person who would like to have my needs met, but that is not the way life is. Life is full of problems and injustices and hardships, but what counts is the personal viewpoint, how we look at these. I had to learn to change many of my interpretations of life, or they would have driven me right off the face of the earth.

I think another reality I had to accept was how little control I had over my life. This was hard for me. I thought I had control over impossible situations. I had to learn to relax and not rage, I did not like it either. I had to learn I had no control over death, which maddened me. I always thought I could kill myself, at least the bad parts of me, but secretly I thought the real me would go on living. I had to accept that I was not two identities, that if I destroyed the part of me I disliked, all of me would follow.

I had to identify and separate my emotions in therapy, and this was most difficult. Fear and anger were difficult for me to accept as normal reactions because I thought these feelings were bad emotions that made me a bad person. These emotions were not tightly locked up either. They came out, but in violent temper tantrums, when I was near exploding, over situations which did not merit the anxiety I was expressing. I had to learn to use my emotions correctly. Damn! Damn! Damn! Also, part of therapy I did not like was being directed or guided back as a small child into thinking healthfully again. I never liked being told what to do, or to be shown anything. Many times I thought my therapist didn't know what he was talking about because he was not burned, so he could not possibly understand what I was going through.

He kept trying, and I believe in time he had complete empathy with me and my troubles. He kept telling me I would get well, and I insisted I would not get well. He said I was good, but I did things to make him think I was bad. He was in many ways like the surgeon who told me I would grow skin to cover my body, but now he said I had to grow another type of covering; this time it was respect for myself. Just as I was terrified to believe the surgeon, I was too frightened to believe my therapist. After denying so much, I had to admit they were both right, and I was wrong!

When the therapy became very painful, and the way I was traveling became too monumental to bear, and I could not stand any more pressure, I turned to doing completely fun-filled things. Once I bought fifty-five pounds of jelly beans and left them around the university in gallon containers to create fun and color. When I was suffering the worst, I always needed color around me.

I had to learn not to let other people's opinions upset me and deprive me of my good self-regard. I had to learn to trust myself. This was very difficult, because if I did that I would have no one to blame for my mistakes. I had to learn to love myself and this seemed to me both selfish and disgusting. I always believed that you had to give till it hurt, and sometimes I gave so much that the pain was extraordinary. There were many times my suffering came from situations of my own creation, and then I did not know when or how to end these painful problems. I never wanted endings, and I dreaded even when a school term came to an end. I was to discover I had subconsciously associated endings with death, and I believed that if a situation never ended, it would always be there and no loss would be suffered.

I did lots of living in the future and past, and tried not to think about the present. Once, I remember being cornered into the living present by the therapist, I could not face up to it, but kept speaking away from that moment, insisting on speaking about what I would do next, or what happened in the past. I simply could not face the present moment and its pain.

CHAPTER XXIX

My "quick eleven weeks" of therapy stretched on and on through Christmas, Easter, the Fourth of July. I was still battling against changing, or changing so gradually I began to think I was completely hopeless and that I would never be guilt-free again. I stubbornly insisted I was right and the therapist was wrong. I often asked myself, "Why, if changing is so painful, do I persist in wanting to change myself and cause more pain?" I could not change the past and all that happened to me, so what was I supposed to change after all. I was attempting the impossible. I would always be scarred and people would always ask me hurtful questions about my appearance. I was a monster who made people recall bad experiences, and that I could never change. Then slowly, an amazing revelation came to me. I learned I could still be crazy and not suicidal, that being slightly zany was okay, but depression was a drag, and a copout from living. I actually did not know when the change started. I could not see any change, but I kept trying.

The summer of my second year of therapy, my psychologist told me he was leaving the university and I was confronted with having to make a new start with another therapist. I did not like ending one relationship and beginning anew with another therapist. This revived my superstition that everything I grew to trust would eventually be taken away, and I underwent a setback in my therapy. I became certain I would never get well. As reluctant as I had been to work with the first therapist, I was certain the second therapist would be even more difficult. However, I discovered I had advanced to the point where I could change and work with another person. So, I learned another lesson; change did not always bring loss and this time it represented growth upwards and out of my depressed state of mind.

A serious depression is hard to fight because it submerges its victim and almost smothers the whole existence. Depression robs its victim of all hope of ever changing. I felt very ancient during most of my therapy. It drained me of much of my energy. Often, after some sessions, I dragged myself home and sat in a tub of water, trying to revive enough to pull myself together.

During this time I had a few friends who were also in therapy, so I wasn't too forlorn. There were other times when I was angry that I was in therapy and at my other friends who were not. Why did I have to change, when they were not perfect either? Why were they able to function and enjoy the many parts of life I was missing? I was often jealous and felt old and tired, and I suffered from another type of pain that was as serious as a burn. There was a greater infection threatening to take over my mind and destroy me totally, and there was no medicine to help me fight against this insidious specter—depression. Depression is a serious illness and yet often it is ignored and not treated as such. In fact, it has become ever more taboo in conversation than death. It is a time when, depleted of energy and with no enthusiasm for living, the victim of depression must carry on the long fight to throw off a heavy pall to get well.

I was always suspicious of therapy as being too self-serving, that it was vain to waste all that time trying to concentrate upon making myself well, especially when so many of my feelings had been squashed until I had forgotten what being "well" meant.

I often refused to admit I was growing and feeling better

because I was so afraid that by merely doing so all the good feelings I was acquiring would be suddenly taken away from me. When therapy became too painful, I insisted I was well enough to discontinue the sessions, but when I actually started to feel well, I grew calmer and could see that some of my protestations were foolish. I used to think I was being tested and if I got too comfortable within myself something awful would surely happen to cause me more pain.

Now, when I think back on those years when I was depressed and wanted to destroy myself, I am amazed that I inflicted so much misery upon myself. I was letting all my anger turn inward, and I almost destroyed myself in the process. It was a long, dreadful, and intricate exercise to recover from disaster, but it was all necessary for me to become whole again, and to understand that while my situation was unique, my suffering was universal. All people have to mourn losses, and I was but one of many people who survived disaster, only to feel guilty that they were spared. Also, burns are a very difficult health problem and there still is little understanding of the needs of the post burn patient. We are very much like a do-it-yourself project.

In my junior year of college, I decided to return to Chicago. I planned to continue therapy because I was not completely self-assured, and occasionally had reoccurrences of contemplating suicide. I would go to bed and then have dreams of hearing my own voice say, "I'll never get well." I was my own worst enemy during these troubled nights when I woke up screaming and wanting to fly out the window down into the streets, trying to escape from myself. Oh, the terror of being unfriendly with myself, the monster within me was so many times bigger than I!

I was in therapy almost three years, and I certainly grew skin faster that I grew my mind back together, and this aggravated the hell out of me; it was so time consuming. The longer I worked the more I thought I was never going to recover. For a person who talks as much and as fast as I do, therapy should have been lots quicker, but many times when I was talking and joking, I was masquerading my true feelings.

That is what therapy is: the vocalizing of your emotions until you learn to accept them without guilt. The emotions and thoughts that for many years I pushed aside lay buried deep down in my subconscious, and I was afraid to dredge them up into the light of reality. Then I discovered I wasn't crazy, I was just suppressing my feelings, and I could not live life fully riddled by fears. I learned that emotions are for sharing and this sharing might involve risk, but risk can lead to growth or even loss. Risk is okay.

Shopping for a therapist can be like shopping for produce in a large, chain, grocery store. You can see the produce through the clear plastic wrap, but you cannot pinch or touch the fruit and must have faith that when you open up your package at home, you won't find a soft spot. I had a few therapists who had a few soft spots or were people I could not work with, but I had been fortunate in finding two therapists at the University in Carbondale that I could work with.

Learning to change was one of the healthier parts of my therapy. I knew I was progressing when I started learning to change gears in order to better myself. I was still having lots of anxiety about pleasing everybody in the world. I was scared that if I wasn't nice, people would hurt me because of my scars, so I had to be very self sacrificing. I had to learn that my scars were not that important to many people and this confused me. I found I had to please only myself and not worry about other people's opinions, because no one can please everybody all the time, or even half of the time.

This was also when I started to let people like me and not be afraid that when people knew me better, they would dissolve before my eyes again. I became more comfortable with people. I learned to deal with endings which are not the worst part of life. I still do not enjoy endings, but I can recover from them more quickly. I can now say I am a good person. I am not scared that I am evil just because someone disagrees with me. I am not afraid to have good feelings about myself now.

Sometimes when I get depressed, I panic because I think it will last for years and years, but I have learned that occasional depression is normal, and I can overcome it when I can identify what it is that is upsetting me.

CHAPTER XXX

I worked eight more months in therapy and found I was functioning better. Many of the voices I heard in dreams and voices I heard during the day were quieting down, and I stopped turning within myself. I continued wondering about the dead, and then I started to wonder about the others who survived, what they were doing and thinking. I had kept in touch with some classmates from elementary and high school, and we started to plan a reunion party for the 1959 eighth grade graduation class from Our Lady of the Angels.

It was not difficult to locate the majority of our class. In about two month's time, we had located eighty-seven members, and we were lacking only thirteen names. It was fun calling up someone I'd spent my childhood with saying, "Let's get together." Whenever we called one person we would find another and another. We talked about the times we had in school. Another girl was helping me plan the party and everyone asked her how I looked or they would ask me about others who were injured. We spoke in softened tones, about a time that affected our whole lives, the fire was still vivid in our minds.

I was surprised when classmates said they worried about those of us who were injured because they knew of troubles we were having. It was the first time I ever spoke to anyone in my class about the day the fire swept through the north wing of the school. It felt good to be able to have this companionship and to actually talk about what happened and not be afraid to share feelings with each other. There were some people who were still frightened and hesitant to speak of the disaster.

On the fifteenth anniversary of the fire, December 1, 1973, I went to the cemetery, and I was deeply touched at the sight of all the floral pieces upon the rows of graves. The sprays of red and white plastic carnations, with ceramic Sacred Heart statues balanced in the center of the arrangement, had streamers blowing in the wind with gold punched out lettering stating, "You are not forgotten, Love Mom and Dad." After all thoseyears most of the little graves were remembered by colorful wreaths, but some graves stood undecorated. Perhaps it was too difficult for someone to come to the gravesite, or perhaps after all thoseyears, Mom and Dad were also dead.

I looked at the large stone monument and read the list of names and the places where some of the children were buried in private family plots. I shook my head and a chill ran through my body. I stood thinking about how I had changed, and what all those little children missed of life, all that I still did not understand.

I always bring fresh flowers when I visit the cemetery, and I try to make them as colorful and as alive as the children were. I know plastic flowers are more sensible in the cold of December. Fresh flowers will not last more than a few minutes in the harsh elements with the wind blowing, the snow on the ground, but I want something alive for the children. The flowers in many ways are like the children, only here for a short time, but full of purpose. I always choose the young, fresh, gay flowers like daisies, poppies and baby sweetheart roses. I include Christmas pine, as if to share the Christmas tree they all missed. They died so close to Christmas and I feel they should have a Christmas tree.

I am certainly sentimental and I place flowers on each of the graves where I knew the child, and there are many. The remaining flowers I usually put on the big monument, but in 1973, I suddenly stepped back and found myself standing directly by Helen's grave. In a firm, quiet voice, I said, "Helen, I am sorry you died, but I cannot be sorry anymore that I lived. I have to be alive."

That evening there was a fifteenth anniversary Mass which was not crowded. Most of the people there did not even know the heartache of Our Lady of the Angels, that still lingered on for so many. There were newsmen, taking pictures, asking the silly questions newsmen always think are necessary, but this no longer upsets me in the church where the aisle no longer challenges me. I could walk, but I still heard whispers, "There's a fire victim" and it was okay. I have learned who I am, and I am far more than just a fire victim.

The Mass started. This sermon was by a new priest who spoke very differently and did not talk about how brave our Blessed Mother was at the foot of the cross when she lost her only Son. This priest said something I had never heard before. He spoke of not knowing the heartache, and the loss, and the suffering we all felt, and he said it must be a time for tears and crying even now. Finally, we were being allowed to cry. How wonderful! The audience was very small, and I thought how nice it would have been if others not in church that evening could have received the long overdue permission to weep.

That night something very curious happened. I went to bed and there were no voices, there were no more bad dreams. I was released to become Michele. I did not have to mourn anymore, my job was done. At first I felt lonely and strange without the presence of the dead around me, but the strangeness did not last long, and I felt very good. I still do not know if it was all the years of therapy, or talking with other classmates and verbalizing for the first time, or receiving permission to cry, but Michele McBride wanted to be alive.

The following February, we had our class reunion. I think it was the quickest reunion in history; we spent only three months planning it. I remember walking into the banquet room and the ceiling was leaking water. "Good grief," I thought, "we have been through hell, and tonight we will go through highwater. Perhaps calling this group together was not the best idea."

The reunion was a success. It was held in a Chinese restaurant for a group of Italian, Polish, German and Irish people. In many ways it was a costume party with everyone trying to guess who everyone else was. Something none of us expected happened that night—we were all especially happy to see each other alive and it felt wonderful.

Part Eight

Confessions of a Fallen Angel

CHAPTER XXXI

Being a "burn" is not a part-time hobby, it is an entire lifetime full of adjustment. Being a burn is to be immensely lonely; it keeps loved ones at a distance and makes strangers curious enough to ask hurtful questions. Being burnt is never in the past tense, which makes it a chronic health problem. For post burn patients there is little understanding and help. Even burn specialists do not fully understand the problems that often arise from the keloid scar tissue that forms after a serious burn heals.

Post burns can affect the body's every function day and night. The circulation, the muscle, the skeleton, the skin, and even the ego are in constant turmoil.

The care we burns must exercise to protect ourselves is tremendous and must be constant. When I stop to analyze all I must do daily to take care of myself and to conceal my health problems, even I am amazed. While all of this has become second nature to me now, it has taken years of hard training to learn how to be a post burn.

There has been virtually nothing written about the post burn victim. I did find some articles on the English post burn written after World War II. Reading these articles made me want to stand up and shout, "Thank goodness the Blitz is over, God Save the Queen and three cheers for our side!" There is only a small amount of material one can read about patients who were discharged from the hospital and left to cope alone. They had to cope alone with the burn as well as with the depression caused by massive scarring. That is all that has been written.

A burn patient becomes woefully frustrated when he tries to overcome his physical and emotional problems but cannot. So the burn patient tends to withdraw, to sit behind closed doors and to stop experiencing life to the fullest. Then fatigue, disfigurement and the pain of crippled joints take precedence. The depression, the despair, the discomfort can change an entire personality.

I still believe there is not a measure for pain. Past pain may be forgotten, but there is always present pain and it is far worse. When I am having an especially uncomfortable day, I do not dredge up the memory of past suffering. That pain is over with and filed away. I have to deal with the pain of the present, and there is no bargaining where pain is concerned.

The physical limitations forced upon me and other post burn patients are endless; every movement, things normal people do automatically, becomes a task. Moving from a chair, getting in and out of a car, such a simple act as standing still is impossible, and holding my hands quietly at my sides are actions I cannot take for granted. I now appreciate the thrill of walking, because I remember how difficult it was not to be able to walk. Walking causes me pain, and I totter or have developed a special gait that gets me where I am going.

I fall frequently and I am angry and terrified with every fall. Because my knees are so badly injured, they have grown weak . . . or is it I am older and the spirit grows weak? Some nights are long and I cannot sleep because I am uncomfortable and my knees and my scars cannot find a resting place. I feel so helpless. I know this is a permanent condition and the darkness of the night intensifies the pain.

Since the long ago fire, I have been afraid to turn on a light after I have gone to bed, as though I am expecting an intruder to be standing at my bedside peering down at me. Perhaps pain is the intruder of my sleep and I am afraid to face it. There are times

I must meet this intruder head on, and then running warm water over my knees, an application of Ben Gay, a glass of wine, or a couple of aspirin will give relief, but I am always aware that such relief is temporary. The nights are long and when I see the light of the rising dawn slip into my room, I feel I have accomplished a job well done. Yet I do not know what I have accomplished, caring for oneself should be a natural act, but for me it is a triumph I relish.

There are many times I don't go quietly about my business of pain, but thrash about angrily. I do not thrash in bed. I am able to rest only on one side, so I thrash about in my head. I get angry, especially when I fall. I get angry when people forget how limited I am, or worse yet, when they do not know. Fortunately, these moments of pain are short, although frequent. Time is triumphant and I remember the moment doesn't matter. It is but a flash and then changes, so I can forget. Sometimes I am quite pleased that people I associate with do not know how difficult walking is, driving is, or standing is for me. Sometimes I tell them how it is with me, and they are stunned and upset but cannot offer advice. Then we all feel helpless, and everyone tries to avoid my pain by pretending it will all go away if we don't mention it.

I alone must stand it, and now I am satisfied that I can. I am happy when sometimes I can share my problems with a few. But I am still annoyed there are many who cannot share with me, many whom I make uncomfortable.

I could scream and I could cry and sometimes I do. What is so wonderful now is not having to deny my problems. I know it is normal to feel guilty about pain, but I do not feel guilty because of the pain that sets me apart. The most important lesson therapy taught me was to take care of myself, not feel guilty, and to make allowances for myself.

During my teenage period, I set rules and regulations for myself that were all very Spartan-like in their rigidity, but now with the wisdom of aging I have discovered I can take care of all my special needs and not think I am pampering myself. The chronic health problems of pain and fatigue and poor circulation are the biggest problems post burns face, not to mention the emotional stress of the disfigurement. Accepting and dealing with these problems does not resolve them, but makes life more enjoyable.

I cannot trust people when they flatter me by saying I am brave and pretty. I always think they are really saying "Oh, she is such a lovely dear, brave person, but scarred beyond belief." It took me years to accept the fact I could be scarred and have a nice personality at the same time. I grieve often about not being accepted, and then I realize sometimes it is I who have not accepted myself.

I have concluded it is drunks and children who keep me honest. I can never deny that my scars show, because a child will always ask what happened. If his mother is within earshot, there will be a big uproar and she will scold and order the child not to ask impolite questions and then make all sorts of apologies to me for her offspring's poor manners.

I sometimes think I am the only burn person out in public because only once did I ever meet someone else who was a burn and was not a victim of the Our Lady of the Angels fire. I was sitting next to a man in one of my college classes, and like everyone else, I was curious about his scars. We started to compare lecture notes every day, and by the end of the term we were acquainted well enough that I could be brash and inquire about his keloid scar. Since he also was one of God's "toasted" people, I felt free to let him know that the manner in which he was injured wasn't important to me, but what I did want to know was if people stopped and asked him what happened to him because of his scars. He told me no one ever asked what happened to him, but if they did, he would punch them in the nose. He asked me how I was injured and I told him. He summed up my experience by saying, "Oh, you're a star." We both laughed because he understood how my being a burn was difficult, and my being the victim of a famous fire had tremendous drawbacks.

CHAPTER XXXII

Burns are a disease without the formality of being medically designated a disease, except in the State of Massachusetts. I wonder if this is because of all the witch burnings there, and if the state is trying to make up for past wrong doings.

Burns are a permanent handicap, and the less visible the handicap, no matter how serious, the more difficult it is to be accepted. This can be maddening for a burn person because the satisfaction and the triumph of recovery is so soon lost after the dismissal from the hospital. All the excitement of growing new skin cannot be seen or appreciated by lay people outside the hospital, and the sight of the plainly seen scars startles people into revulsion and this hurts the post burn person. We're disfigured and the flames that once engulfed our bodies become permanent structures on our bodies. This is why a fire does not die for any burn person and this is why I titled this book, "The Fire That Will Not Die."

Fire consummed one of the largest organs of my body—the skin. Skin is the least appreciated and understood organ and without its protective covering, man would die. Skin has four functions: (1) as a protector of underlying parts from mechanical and radiation injuries and from invasion of foreign substances and organisms; (2) as a sense organ; (3) as a temperature regulator; and (4) as a metabolic organ in the metabolism and storage of fat, and in water and salt metabolism by perspiration. With all these vital functions going for it, skin is the least respected of all organs and the one most taken for granted.

Whenever the skin is thought about, only one ninth of the body's total skin is ever mentioned, and that is the facial skin. We forget this area of skin we spend millions of dollars on every year does not stop at the neck but stretches over the entire body. Whenever a large area of skin is destroyed, the whole mechanism of man's existence is upset.

I have often thought what a great improvement it would be if we humans could be covered with skin that would not burn. I have not exactly prayed for this, but I have noted it in my ideas to God on how He could improve the world—by adding one little feature, and that would be to make human skin flame proof.

Skin consists of a beautiful network of many cells, intermingling within a definite pattern, forming a masterful chain of elasticity covering the body and protecting it from infection. Skin warms us. Skin is also the monitor of our mortality. It will sag, it will wrinkle, and by these signs we can condition ourselves for the final days of life. Skin is the method by which people instantly judge us. We might have hearts of gold, but people do not see that. The color of our skin may perniciously divide us. Skin is beauty. Skin is life.

The two layers of skin are composed of the outer layer, the epidermis, and the lower layer, called dermis. Intense heat, no matter of what type: flame, electric, explosive, or steam, destroys all the layers of skin exposing muscle, nerves and bone. The regular pattern of cells is destroyed, the uniformity is gone and the irregular, mysterious pattern of keloid scarring is formed. A deep burn destroys oil glands, sweat glands, hair follicles, nerve endings, and blood vessels. When the regular pattern of cells is destroyed, keloid scars form. The fibrous tissue overgrowth occurring in scars forms a thick sensitive mass—but medical science has never understood why this is so.

When large areas of flesh are destroyed, and so many functions of the body are damaged, the constant need for skin replacement becomes mandatory. Perhaps I am overstressing these facts because I saw my skin disappear so rapidly that I went from a little girl to an instant old lady in ten seconds. That which takes most people a lifetime to accomplish, I did in a brief moment. Because of this, I do not fear aging. I am not shocked to look in the mirror and see my skin growing older. That shock occurred many years ago, when I had the vitality of youth to accept it.

The post burn person is physically limited in many ways and his body is subjected constantly to strange sensations in various areas. At times I actually believe I must be glowing in the dark because of the various sensations flowing through my body all at one time—sometimes a numbness, other times a sharp, electrical twinge. There are times I am confronted with certain tensions in my work, and the skin on my legs goes into a dramatic change. There are throbbing pains for no apparent reason I have ever been able to figure out. I cannot simply excuse myself and leave a room because my legs hurt and I want to elevate them.

I am an ace bandage junkie and I always need one wrapped around my knees because I fall frequently. I feel that I know a football star's leg and knee problems without ever having been in a game.

My legs feel better when they are propped up, and in the comfort of my home this is fine, but outside I cannot have this luxury.

I cannot stand still, and this presents problems from the moment I get up to bedtime. Do you know how it is to waltz while brushing your teeth, or gluing on false eye lashes? This is not a big problem, but it is an everyday occurrence I must deal with. Also, I must keep moving in the kitchen while cooking and washing dishes or my legs will turn a deep purplish black and start itching.

Once when I was giving a dinner party, I scolded myself because I was always resting in the living room with my legs propped up on a hassock. When I started to peel vegetables on my lap while keeping my legs propped up, I decided I was just plain lazy. I discussed my problems of homemaking with a girl friend, who was recovering from serious surgery, and she said her energy level was so lowered she found it necessary to rest more often during the day to make housework easier.

I had to convince myself this was true, and stop saying I was lazy. I had to stop browbeating myself for not being able to stand for any length of time. I learned not to think of myself as a lazy, self-indulgent person, and that taking care of myself is not being neurotic.

I have to exercise my stiff knees regularly and this is a real annoyance to me. I think exercising should be reserved for health nuts. I have read a great deal about exercise and I have come to the conclusion no one really likes to exercise, so why should I be any different? If it hurts to get in and out of a chair, I rebel at having to go through the regimentation of exercise because I have to strengthen weak muscles or my knees will collapse more frequently. Exercise is healthy for everyone and why should I be the exception. I have calmed down and I have trained myself not to exercise because I am burned and have a special problem, but to exercise because it is good for everyone. I wish I could honestly say I jump out of bed and start every day doing twenty minutes of regular exercise which I should do, but I don't always do. I have increased my exercise habit more and I must confess I feel better when I exercise, but I don't like it.

Diet is another requirement for the health of a post burn person. I certainly do like to eat, and when I reached the age of twenty-five I found myself overweight, and I had to lose twenty-five pounds. I did this as quickly as possible and nearly played havoc with the keloid scars on my legs where the skin is transparent and thin and does not have normal elasticity.

For some unknown reason, steady, even pressure upon the burn locations helps reduce some of the distressing sensations that continue to plague me. I like to sleep under nine quilts, not for warmth, but because the scar tissue is so fragile and vulnerable the weight of the bedcovers feels comfortable. The one exception is the inside of my left leg that cannot stand the slightest pressure on it because of the insertion of a shunt during those spells of renal failure back in the hospital.

The scars on my back are a mixture of skin grafts, keloid scars and donor sites. The donor sites become chafed during the summer heat, and the thick keloid scarring starts feeling sticky or pulling, like maple syrup running down my back. They do not ooze, but that is the sensation I feel, particularly when I am in an air conditioned car or room. The contrast between the pulling of the thick keloid scars on my hands and back, and the tightening of the skin grafts on my legs, is very peculiar and hard to describe. It seems to me that every area of my body responds differently to temperature, dryness, light, and body movements, until I feel like a crazy quilt.

When one mass of scarring starts to react to a certain situation, I sometimes think, "Oh, not now," but I can honestly say I have never cancelled an engagement because of the complications of this scarring.

I know from experience it is better to keep busy than to dwell on the problem; being active is so much better than sitting idle. Sitting can cause problems because I have a burn scar on my buttocks, and in the summer that burn will weep, and that is when I certainly know I have been sitting too long.

Some grafts on my legs always feel numb and very taut. I imagine this is similar to the people who suffer from edema. It is difficult when I break out in rashes because if I scratch, the burn area will bleed.

Whenever hair follicles and oil glands and sweat glands have been destroyed, artificial lubrication must constantly be applied to the injured skin. Like many post burns who find great relief by splashing water over the affected area, I like water and at least twice a day wet myself down. It should be noted that burns should use tepid water and it is best not to towel dry the burnt area.

All doctors prescribe swimming for a post burn patient, but I know many burns are too self-conscious about their scars, and cannot don a bathing suit. It took me a very long time to be able to go swimming. I get up my courage, all I do is make a wild dash to the pool—jump in, swim, and come out, then hurriedly put on a long concealing robe. I also have a distinct advantage when I swim because I do not wear my contact lenses or glasses, so I cannot see anybody stare at me. I am so nearsighted that when I take off my glasses, I am the only one left in the world. Another grave problem is my poor blood circulation that cuts off oxygen to the burn area and makes it look sallow. This discoloration changes from time to time and I do not know if it goes from good to bad, or bad to good. I have marbled skin that looks like an Easter egg and is as fragile as an egg shell. It often breaks down and results in many health problems.

I cannot tell if all of this skin tenderness is painful or just annoying, like a fly buzzing around me that I can't wave away. There are times I sit and try to brush aside these annoyances and tell myself to ignore these silly sensations that are really nothing but old scar tissue acting up. I guess what really gets to me is the fact I have no control over these pesky sensations; they can strike me at any time and without warning.

CHAPTER XXXIII

Proper grooming and dressing after I became a burn was an art that took me a long time to master, and I could accomplish this only when I began to like myself again. To learn to love myself was a hard lesson.

Dressing to camouflage my scars was an art all by itself, and I have learned by trial and error what is suitable for me. It is fortunate I come from a family where good grooming is a part of life and I know the struggle of properly clothing the body. I have figure faults like other women, but mine are more pronounced and I have to conceal my arms, back and legs.

There is often a flurry of consternation when the saleslady comes into a dressing room and sees my scars and asks questions. I stand there thinking, "No, you really would not enjoy hearing what happened." If I say "the Our Lady of the Angels Fire," I know I am going to hear that bromide about my having been a brave girl, when all I want to do is get the right cocktail dress or a pair of jeans—just some damned outfit to cover the body. I insist I have the right to all the frustrations of getting into a skirt and top that fits around my hips and bust like normal women. I do not want to be told how brave I am, when after all, I am only shopping.

I'm bored about being told I am brave. If I ever broke down

and told those clerks all about the Our Lady of the Angels fire, I'd have to hold a convention in a dressing room. Sales persons mean well but they seem to congregate from nowhere to ask questions while I am standing in my underwear wishing they would leave or at least get busy and find me something to buy.

I cannot overemphasize the trouble the post burn has to endure with the keloid scars. The keloid scar causes problems that are ever present and ever changing, like a chameleon.

Somehow, someway, while I was a patient at St. Anne's I accepted the fact that I could always be scarred, that the scars would fade somewhat, and that I would always have a certain amount of disfigurement. I am sure I did not say disfigurement, but called it a scar. I never expected all the peculiar sensations from these scars that would follow me throughout life. I never anticipated the skin grafts on the left side of my face would pull and feel taut. The softening and fading out of the keloid scar on my left cheek has taken place during the years, but cold weather causes that scar to turn vivid shades of rose color and to become very hard and then a tugging sensation takes place. One graft on my face stings and twitches but I can never pinpoint just when this might happen since all weather changes seem to affect it. There were many first degree burns on my face and this skin puffs up in dry heat.

The keloid scar on my nose can easily be concealed with makeup and made to look like normal skin. Only I can see the faint line of the huge scab that fell off my nose in the hospital with a gush of blood. I cried when this happened because I thought my nose had fallen off. Many years later I was relieved when a glance in the mirror showed that I had retained my profile and still had my little nose. The softening of all the scars has made the application of makeup easier, and I am not allergic to most makeup, which is good.

I think using proper cosmetics is good protection and it has helped me adjust to the reactions to my face that I have been subjected to every day. I studied to learn how to wear makeup not just as a base, but as a coverup. With the softening of scarring on my face and the fading of the vivid scar tissue, I have found it necessary to change my makeup several times to keep up with all the varying changes in skin tone brought about by healing.

I had to decide how to pull this all together. Now I use one basic color on my face, but by skillful eye makeup I draw attention away from my face to my eyes, and I have learned to use my eyes to the best advantage, and even flirt with them. When speaking to people I try to look directly into their eyes, and most of the time the compliment is returned and people are attracted to my eyes and not to my scarred face.

I had to find a different way of combing my hair to conceal the effects of head surgery that left me with a new hairline. The hair on my left side is about one and a half inches higher than the right side, making it impossible for me to wear my hair off my face or pulled straight back. This is not a great problem, but I do have to consider this when I have my hair cut. Also, because of the scarring, I started parting my hair on the right side during the 1960's when this was not done. Hairdressers said, "No, only men part their hair on the right side."

I have to be very careful about the burn scar on top of my skull. If I do not keep that burn covered during the summer time, it will become sunburned and swell up. Thick bangs hide all the stitches and lumps on my forehead, which are leftovers from the skull fracture. Some Eastern headreader would have a field day rubbing all the lumps on my head to tell my fortune.

The surgical re-arrangement of my hairline has made me feel lopsided. Consequently, I am very careful to keep my hair always in the same style which is cut with the hair falling around my face. This facial skin is very sensitive and dry and the hair falling on either side of my head acts as a protective veiling from further injury. There are times I would like to pile my hair on top of my head, like old washer women, and go about my business. However, I am too vain, and I could not endure the reactions of people when they saw the crooked hairline and lumpy forehead.

Also, because of the placement of a skin graft, I lost half and eyebrow, so I do not make up my eyebrows; I just leave them alone.

There are many keloid scars that have softened through the years and are not noticeable to anyone but me. These took about ten years to soften and become flexible, and have smoothed out nicely except for a certain waxiness which still is in the process of changing. When I wash my makeup off, I have to take special care to see that all the lower crevices in the scar tissue get thoroughly cleansed to prevent any infection or breaking out. I wear lots of colored eye makeup to distract from the sallow skin tones of my face caused by loss of much pigmentation. In many respects, despite all of the difficulties I have undergone the last twenty years, there has been a remarkable change for the better in my appearance.

When I turned tweny-five, as I have already mentioned, I suddenly realized I had let myself put on too much weight. Food has always been a great delight to me and I had to cut down right away. I wasn't prepared to see the changes weight loss would make in my skin, but after seeing how loose this flesh hangs after dieting, I must accept the fact that if I jiggle my weight around, my skin will take a very big toll. On the other hand, the skin on my legs is abnormally tight in some places and cannot withstand the burden of excessive weight. Also, with weak knees, I have to keep the weight down and I might as well stop feeling like a martyr doing so.

This martyr complex is a part of my depression I have almost conquered. There are days I am dressing and look at my body and see all the mobility I have obtained and I rejoice that I can get in and out of a chair and even a bath tub, but there are other days when my scars shock me and I realize how disfigured my body is. There are the days when my eyes well up with tears and I lament the loss of my young healthy body.

The art of good grooming and proper dressing is very important to a recovered burn patient. I wish I could counsel and impress upon each and every one of them that before any attempt at self beautification can even take place, the burn must learn to love himself again and again. One time I told a boyfriend that if I wasn't scarred, I would be an exhibitionist. He assured me I am an exhibitionist nonetheless. Fantasies of possessing a healthy body dance in my head some days, and then I believe that everybody else in the world has a center-fold type of body. One trip to the beach proves this idea false, but I look around and see tanned, non-burned bodies and know that I can never bask in the sun. First of all, sunburn is not for already burned skin, and I have been cautioned that too much sun might cause skin cancer.

Fire maims victims for life. It is very difficult for a post burn to forget the trauma of the bad experience when their very physical being has been interrupted for life. Burns heal but their disfigurement lingers. The pain of disfigurement is hard to deal with in a world where physical perfection and beauty are thought necessary before happiness and contentment can be obtained. In our culture it is demanded that we all have perfect teeth, perfectly formed breasts, cute little noses, etc., and this is supposed to insure perpetual happiness and freedom from insecurity. A perfect image is supposed to attain for us more than our constitutional rights. However, we badly burned persons have no access to this perfect image that is supposed to grant all the happiness there is in the world. Our self image is destroyed, putting our best foot forward is impossible, and yet we must go on and live in a world in which perfection is sought and thought to be the ultimate goal.

Society grows weary with our scar-marked ugliness and our pain. To have survived a holocaust, then endured the pain of treatments, to have clung tenaciously to life when there was so little hope and so much misery, then to be socially rejected because of our looks, is heartbreaking. It is beyond endurance.

As I mentioned before, I am always cold; the chillier I get the more severely I hurt, until sometimes I panic with the intensity of the pain. This panic scares me more than the pain, so I have contrived ways to keep warm. One way I do is to carry a thermos of coffee with me to warm me. Loud, colorful knee socks help stave off the cold, and I try to find the brightest knee socks I can because they remind people of funny things. I have one pair of red, yellow and orange socks and people's comments on them range from their being from the Land of Oz to Raggedy Andy. No one is aware my socks aren't for silliness, they are necessary to cover burned legs and keep me warm.

If people ask me what helped me survive as a burn, I can say a color wheel, my red, comfort blanket, and my happy, riotously loud socks. I believe there is more truth to that statement than if I told people my recovery was a super human feat. It has been fun to see just how much I can get away with. Perhaps if I dressed quieter, people would not notice me, but I like to think it is my choice of wardrobe people are staring at and not the scarring. My happy garb has helped me survive the pain and sometimes almost insurmountable rigors of entering the mainstream of society, and I'd like to share this secret with other burn victims.

In this day of psychological and scientific advancement, I do not understand why someone has not come forth with a counselling program specifically designed to assist the post burn in solving his/her many problems.

There needs to be such a program for re-entry of the burn into the mainstream of life and for other disaster victims.

The very concept of a burn's self image is to conceal the vivid, torrid, inflammed, marroon scar tissue that clings over our bodies like hot melted wax poured from an invisible caldron. It distorts our faces, our bodies, our appendages. It pulls our bodies tight where they should be smooth, it limits our motor control, it raises our anxieties, it scars us, it makes us hostile, we are unprotected with the beauty of skin, our looks repulse others, we are hurt.

Sadness wraps our existence, making our recoveries imperfect and incomplete, and at times unwanted. Our presence pains others and people will stay away from our injuries by avoiding us;

our pain cannot be shared; we must be silent so as not to offend more than necessary. Insecurities crop up and we start living in a world of self doubt though we are encouraged by the commands from the medical practice "give it time." Scars do change, but never are they completely erased. We are cautious, moving already in a restricted environment: the body. We are afraid we are an outcast; people are afraid to say anything to us and avoid us. We withdraw, we are discouraged. We forget many of the human qualities of life because we are so wrapped up in scar tissue.

We forget patience and understanding. We want support, but we do not know how difficult support is to give. We have experienced so many different aspects of life that we are living in different ranges of experience. Burn patients have been through the trauma of dying and with recovery we forget that there are many afraid of us because of the trials we have experienced. We not only look different but we have different attitudes. We have had the opportunity to arrange our values, the time to question our identities, and to search for the truths of life. We have all prayed for help, we have cursed the Almighty for letting us suffer, we have viewed life from another standpoint, and we find re-entry into the world very hard. In many aspects we have gained strengths, but we are weakened by our physical handicaps. The loss of good self appearance, combined with chronic pain and fatigue, make for a loss of self-esteem.

To learn to accept these problems is not the first step toward recovery. Rather, to learn about these problems and to be able to identify each one is the first step. To be able to find ways we can make ourselves comfortable is necessary. There are many little adjustments we are forced to make, that are individualistic, and not necessarily a universal problem of burns. It all depends on the extent of the injuries, what areas of the body and how deep the burns are. If you are burned in an area involving joints, you will suffer weakened joints everyday. The caring of your body becomes a necessity and you have to accept the fact that you will never have non-injured limbs. But you must take care of what is left, you must accept the loss and prepare yourself for what your limitations are. I know how I am limited when I have trouble walking on uneven terrain, cannot run or skip, cannot hold packages down at my sides, all because of poor circulation. There are many professions requiring leg work, and I cannot consider them. Because I am very mobile, it is easy for people to forget all the things I cannot do. I cannot stand still and this makes waiting in lines more difficult for me. Even standing up at a party, I must move or else my legs will turn black and start hurting very badly. There are times I am cornered in a situation where I stand too long. Then I limp more; the pain becomes more pronounced and I get crabby—crabby for not taking care of myself and crabby I am feeling badly. It is during these times that I have to be very aware I am short tempered and irritable. I have to remind myself it is not a direct action against me that I am hurting, and I have to find a way to ease the situation. I have to be aware and honest with myself when I am feeling especially bad and troubled with my burns, that I do not tend to withdraw and become introspective. I have reasons to become angry, too. I will flair up with anger, and fly off the handle just because I am human, which has nothing to do with my burns. It is hard sometimes to know the difference, and being aware of the problem doesn't always solve the problem.

I think being able to accept the pain is the most important factor in dealing with chronic pain and health problems. Understanding that it is not going away, that there is no magic relief from it, is one step in dealing with pain. Pain is not a punishment, but the result of a serious injury which is healed—and the healing process isn't one hundred percent perfect. It is like breaking a favorite china cup and then gluing it back together; you can see the thickened mucilage harden holding the broken pieces together. You have the choice of letting the cup sit on the shelf or using it with caution and getting enjoyment from it again. Post burns are living people who must now live with caution to protect themselves. The caution does not have to interfere with living, but rather it should be used as a safeguard to go on living. Taking care of ourselves is the best way to proceed with life. I notice when I take good care of myself, I am happier.

I was very demanding of myself, but with experience I have learned to ease myself and not drive myself when I get uncomfortable. I cannot say I do not get angry from my limitations. I am very aware of all the concessions I must make. It is better than denying them.

Part Nine

The Phoenix Program

CHAPTER XXXIV

The Phoenix was a legendary bird that lived in Arabia. According to tradition, the Phoenix consumed itself by fire every 500 years, and a new, young Phoenix rose from its ashes. In the religion of ancient Egypt, the Phoenix represented the sun, which dies at night and is reborn in the morning. Early Christian tradition adopted the Phoenix as a symbol of immortality and resurrection.

I have probably done everything wrong in overcoming the problems of being a post burn, but it was all trial and error on my part, because there was no help, no re-entry to life program to guide me. I am fortunate that I have been able to retain much of my good looks and yet I still feel the effects of discrimination.

Because I am a typical burn victim, and have endured all the pain, the hopelessness, the embarrassment, and the sorrow other burns experienced, I have tried to establish a program for the post burn. The "Phoenix Program," as I call it, is a rehabilitation program to help God's "toasted people" cope with all the problems that daily confront them. Some are medical problems, some are psychological problems, some are grooming problems, some are just hell. I always wonder when someone says to me in amazement, "So you are one of the Our Lady of the Angels fire victims." I feel they believe I have sprouted wings like an angel or that there is still the smell of smoke around me.

I rather think I have other abilities and I am trying to use them to help my fellow burns. I want other burns to learn that life is good and wonderful, and so full of problems that it is exciting. I want to tell them that there are times when I am so impressed with being alive, that I can cry. I will tell them this is not the time when I see a rising sun, or a fresh blade of grass, or the first robin in spring, that it is not patriotic or dramatic. Perhaps it happens on a crowded elevator when I can make someone smile, or perhaps when a friend doesn't say, "You're lucky to be alive," and instead says, "I am happy you are alive." That means so much more to people who have survived life and death situations. I try to make my burn friends understand that people are afraid to say this, just as they are afraid of saying good-bye to the dying. Normal people are timid, too.

I am fortunate that as I grow older, I am not growing toward death. Instead I am growing away from death. I grow away from a time when I did not want to live. I can celebrate all the gray hairs and birthdays, because there was a time I was my own worst enemy and I did not want to live. I am not afraid of death and this makes life fuller and more meaningful for me.

Science has finally caught up with my experience and now is studying people who have clinically died and then have returned to life. Perhaps, I am one of those people. That is the only explanation I have for the intense feelings of being left behind and returning to so much loss after the fire. The peace and total acceptance of self during the most torrid moments of the fire, when I was burning and crashing to the ground, are not the most frightening moments of the ordeal I was to undergo. There is an overwhelming blessedness that I experienced during this time that has been unmatched.

I witnessed a twilight of happiness that is not meant for the living because we would not strive to plan or to dream—man would be lethargic if he knew the contentment of death. Having endured a tramatic experience and having discovered a lasting peace has brought me to terms with death. I would not do anything to hasten death as I once did, but I would not go to extreme measures to avoid death. I would use my energy to share death and make it more meaningful. I know I'm more frightened of injury than I am of death. I carry many fears with me that I will once again see and feel my body be hurt. I envision dropping elevators, falling ceilings, crashing cars and I am nervous about such events. I feel the pain that my body will once again explode. These fears are constantly with me. I remember the extreme pleasure of stepping outside my body and I am not afraid, that one day I shall die. I know I was outside my body. I do not feel sad that I did not die because I learned so very much from the experience. I have a greater knowledge which is rewarding and I have the experience of another world—a world where you gain everything but where you must leave your total self behind.

Burn patients need encouragement to develop good feelings about themselves. As a result of the unexpected trauma of being burned, the long painful treatment and recovery, the complications of the life-death syndrome and the monumental task of learning to care for injuries, the post burn patient needs considerable help to lead a full life.

This dismissal from the hospital is only the beginning step toward recovery for a post burn patient. The unexpected toll can reach into categories far beyond the extent of the injuries. Many fires have caused multiple deaths of family and friends, destroyed homes and careers. The post burn person has limitations which cannot be resolved. The life-style of the burn victim has changed dramatically for not only himself but for all family members who have to realize the full extent of his limitations. It is difficult to realize that we are always in constant turmoil and our scarring is reacting to the ever changing environment. These changes could be seasonal, the air-conditioning, heat, bright lights, dry air, humidity and there is very little that we can do to avoid these environmental conditions that we must constantly take into consideration.

Having the general public educated to know what keloid scar tissue looks like would be beneficial for the post burn. Having people ask us questions about our scars is very monotonous and sometimes embarrassing. The general public needs to have more understanding for burns and more knowledge of the health problems involved with our handicap.

Burn patients need encouragement to develop good habits in caring for themselves, and I especially believe they need to learn the art of make-up and proper dress. I have coached many a burn patient on how to start learning proper grooming, and it is exciting to be able to help them. However, sometimes I feel so helpless, since I understand the problems and I am not able to help more burn victims solve them.

My life situation has been very unique and that is why I have the drive to make the burn a national and universally recognized disease. The Phoenix Program, I hope, will become the support group for burns and their families everywhere, so that they, like the mythical Phoenix bird, can rise again from the fire which continues to burn every day of their lives.

The advancement in burn therapy has made rapid strides in recent years. Ironically, the use of napalm in the Korean "Police Action" and the war in Vietnam has hastened this advancement. So many napalm burned victims came out of these wars that significant new treatments have been developed: salving of burns to reduce infection; temporary skin grafts, first from human donors (including the dead), then from pigs, to speed the healing; a special tub-like bed filled with tiny silicone beads buoyed by compressed air which can help relieve the pain of a bed-ridden patient; the wearing of special pressure suits designed to reduce scar burns; the use of hypnosis to help quell pain; etc.

There are "burn centers" being established all over the world.

THE PHOENIX PROGRAM

Nearly every state in the United States now has a burn center. Three of the world's finest burn centers are located in the United States: The UCLA Medical Center and the Sherman Oaks Medical Center, both in Los Angeles, California, and the Brooke Army Medical Center in San Antonio, Texas (civilian as well as military burn victims are treated here). These three burn centers, and many others, (e.g., Boston Children's Hospital and the Cook County Hospital in Chicago) provide psychiatric counseling for the burned patient.

But what about therapy for post burn patients? Are they still being sentenced to live behind doors once they are discharged from the hospital? This does not have to be. The post burn is discharged alone to cope with re-entry into society and like me, many of them will make many needless mistakes. Some will be able to adjust, but others won't—and they will be the lost ones who have given up crying out for needed help.

I have talked to doctors, plastic surgeons, burn specialists, nurses, social workers, hospital administrators, and make-up consultants in hopes of starting this national rehabilitation program. I have written feature magazine articles about the problems of the post burn and all the results have been:

> Dear Ms. McBride:
> You have a wonderful idea. There is a need for rehabilitating the post burn, but I am sorry we cannot help you at this time. Good luck.

Letters like this are stacked on my desk.

From being told for so many years that I am lucky to be alive, I do not accept the definition of the word "luck" applying to the Phoenix program. It will take more than good fortune to establish the program.

Many times I think I should quit and go into an established profession. I think about giving up the idea. It's an impossible dream—to want to help all people who have been badly burned. But those old feelings from the days at St. Anne's Hospital keep coming back, those wonderful moments of encouragement keep creeping back into my mind. I hear the gentle cooing, "Michele, you will get well, you will recover, you must endure, you must take your medicine, you will grow skin again. Michele, you will use your hands again, you have to give it time, you have to work hard, and it all will happen."

Damn, I think, why me Lord? I went through a disaster and now I have more positive feelings from it. I am lazy. I do not want these feelings. I try to get rid of them and forget them. I say to myself I am not going to inquire any more about establishing the Phoenix Program. Then I wake up in the middle of the night with one more idea, one more person to contact, and it starts all over again.

I have bored my friends talking, planning, discussing my schemes on how I am going to establish a program for those who survive fire. I have worked with other post burn people and it saddens me to hear people say post burns are hostile, over sensitive about their injuries. The public does not have a notion of what it means to be a toasted person. People think burns are in the "past tense." They do not know how difficult it is to be a burn, with all the adjustments that have to be made every day.

I have the hope that one day people will look at my scars and know that the scar tissues I have on my body are a result of burns. Perhaps then they will understand the health problems I have. I do not want sympathy, I want understanding. If the public is educated to understand burns, perhaps the public will then become more fire safety conscious, and do more to prevent burns.

I go out, shouting for the burns to come and help each other. I cannot locate the post burn. I read about them in the papers, hospitals admitting them, but then they disappear. I know burn patients do not dissolve in ashes. So I wonder where they are—and I know they are wrapped in a fire that will not die for them. They are alone, suffering physical and emotional problems; they let guilt and anger and depression rule their lives.

I know I will not rest until "Phoenix" is a program across the country helping all burn victims and their families. I know I will not try just one more time, but as many times as needed to help establish the program.

I have tried to give up the idea, but I hear a fire engine racing down the street and I think about the time in second grade I was taught that every time I heard a fire engine I should make the sign of the cross for the safety of the people in the fire. I remember making the sign of the cross when I heard the fire engines pull up to Our Lady of the Angels School as the room was fast filling with smoke. I prayed then for the safety of the children, but now when I hear the sirens of a fire truck, I just pray that no one suffers long, and I am not quite certain what length of time I have determined as being long.

Burns are in the past, present and future.

I know that right this moment, someplace, somewhere, someone is burning. Somewhere there is another fire that will not die.

Appendix

It Need Not Have Happened

CHAPTER XXXV

Fire! Fire! It is all around us. We live in a burn prone society where we think it will never happen to us. We don't understand the dimensions of fire, and yet it is the most present, destructive threat to our lives. We do more to safeguard our possessions from burglary than we do to safeguard ourselves from fire. Everybody locks their doors to keep away robbers, but do we safeguard our homes, work or school? No, because fire happens to others not to us.

People smoke in bed, they practice poor housekeeping, (which does not necessarily mean dirty housekeeping, but dangerous housekeeping), keep combustibles under stairwells, and do not keep matches away from children. We have outdoor barbecues and inside fireplaces. We use matches and candles and do not use the precaution we should. Fire does not deserve respect, it *demands* respect. There is no building which is flame proof and smoke proof. That is why we must all take the necessary action to prevent fire from becoming our masters No matter how careful we are of our building materials, we must always consider that skin is not flame proof. Wherever you have man, you have the threat of fire. It is that simple, man is not fire proof.

Fire is an angry god who will not tolerate lack of worship; it will flare up at any given moment without notice. It will kill, destroy, maim. It will demolish homes, buildings, possessions.

IT NEED NOT HAVE HAPPENED

We cannot control fire, but we can learn greater cautions concerning fire. What is so sad is that most burns can be avoided. The human error involved is startling because practically every burn can be avoided, whether it happens in a car accident, or a backyard barbecue. Burns need not happen.

Our Lady of the Angels fire was no exception. I discovered this while researching my book. I investigated the causes of this fire.

I began by reading what the "1959 National Fire Protection Association Report" had to say about Our Lady of the Angels school fire. According to this official report, the cause of this fire is still undetermined. After twenty years, the Chicago Fire Department still has the case open for investigation and the official coroner's report declared the cause of the fire "undetermined." Fire inspectors pronounced the school fire-safe just a few short weeks before the holocaust occurred. However, there were many existing violations of the fire and building codes.

I read other reports and I interviewed many involved people. I discovered new, and reaffirmed many disconcerting facts, until I began to think that there was a downright, deliberate, planned coverup to confuse the people of the parish.

Even the blue ribbon jury which was selected to investigate this fire was warned not to get involved with religion, and not to ask too many questions of the nuns. The investigation of the fire was kept on a high, dramatic, emotional plane, but the cause of this fire was never disclosed. Mothers of dead students were put on witness stands to testify that fire safety should have been enforced, but their pleas were ignored. Everyone knows that the Chicago fire was started when Mrs. O'Leary's cow tipped over a lantern, so the song says, but the cause of Our Lady of the Angels fire remains undetermined. I was also disturbed to hear there are people who claim that it was the work of a known arsonist who remains at large.

In my child's mind, I always thought that it was an electrical fire, because of the ceiling crashing down, but I know now that this was not the cause of the fire. At first, I was not too interested in how the fire started. I dwelt more on why it happened and was permitted to get out of control. I am afraid I found there are no clear cut answers for either question, but I did discover far more human error involved than even I suspected at the time of the fire.

Reading report after report written by investigators who studied this fire, I found there was evidence of considerable delay in transmitting the fire alarm. The facts about this tragic delay are a matter of record.

The more I researched the human error factors, the more certain I became that the Chicago Fire Department bungled deplorably and failed on every score while putting out the Our Lady of the Angels fire.

Twice during this disaster, I tried to wait for the firemen's ladders, but they didn't quite reach the window and so it was either jump or burn to death for me. I know children who called for a ladder to be put up at their window and had to wait for the firemen to decide which window to place the ladder.

If it had not been for the fast action of the eighth grade boy, more would have perished, and it was a layman and a priest who rescued most of those who lived from my class. Most of those who escaped the fire alive did so on their own accord. Just as I jumped out of a window before burning to death, many others of the injured had to jump in order to save their own lives. Out of the three hundred students who were confined to the north wing, more than half of them either burned to death or risked jumping. Thus the fire department can claim credit for helping less than one third of the children who were trapped in the building.

It was a full fledged disaster. Why wasn't the fire department adequately prepared to handle the evacuation of this building, particularly when the school was in the fire department district? In finding answers to some questions I was encountering new questions at the same time.

Regrettably, the answers learned from this fire were but a repetition of lessons which should have been learned from fires in previous years. If only someone had heeded the warnings and brought the school building into conformity with the Building Code - especially the requirements governing exits, many lives of little children would have been saved.

Shown below is the report from the January, 1959 National Fire Protection Association Journal on Our Lady of the Angels fire.

It Need Not Have Happened

Again it must be written that the lessons learned from this fire repeat lessons learned in years gone by. Again it must be said that conformity to the provisions of the Building Exits Code[1] would have prevented this disaster. Again it must be wondered how much longer it will be before the lessons so tragically brought home repeatedly by school disasters are applied to all schools.

The loss of life in this fire was primarily due to *inadequate exit facilities*[2] as discussed in the following section on exits. This is a basic principle of life safety from fire. Five other weaknesses in the fire safety of the building also made major contributions to this holocaust.

Exits

Basically, the adequacy of exits is determined by proper enclosure, by provision of at least two ways out remote from each other, and by sufficient exit capacity so that all occupants can leave the building promptly. In none of these respects were the exit facilities of Our Lady of the Angels School adequate.

Preordinace Buildings

In 1949 the city of Chicago adopted a Municipal Code which

incorporated all the major features necessary for life safety from fire in buildings, including enclosure of stairways in schools. However, important provisions of this code, including enclosure of exits, did not apply to the north and south wings of Our Lady of the Angels School and to other schools in existence when this code was adopted. In other words the substandard exits in all but the annex (built in 1953) were of preordinance vintage, hence the non-retroactive law did not apply. Why the annex stairs were not enclosed to comply with the law is not known.

Exit Enclosure

The Building Exits Code requires that all stairways in school buildings be enclosed so that in case of fire the occupants can escape without danger from fire, smoke, fumes and resulting panic. The stairways in Our Lady of the Angels School were open except the two in the front of the north wing. These stairs were enclosed at the second story level by substandard doors which were blocked open at the time of the fire.
It would have been fairly simple and inexpensive to enclose all stairways properly. If this had been done the 93 lives lost in this fire would have been spared.

Exit Capacity

The 9½ units of exit width from the second story of this building were sufficient to permit 570 people to reach the ground in 1 minute and 20 seconds, according to exit capacity requirements set forth in the Building Exits Code and elsewhere. It is conservatively estimated that there were 610 people on the second floor of the building when the fire occurred. A closer examination of the distribution of exit capacity shows a striking contrast between the exit provisions for the north wing as opposed to those for the rest of the building.

In the north wing, the seven exit units were more than adequate *in capacity* to handle the 329 people on the second floor. In the annex and south wing, however, there were only 2½ exit units available to accommodate the 281 people believed to be on the second floors of these two sections. Two-and-one-half units of exit width are adequate to evacuate only 150 people in 1 minute and 20 seconds. The inability of the 2½ exit units to handle the 281 people in the prescribed time was demonstrated by the fact that the average time for the evacuation of the second floor in eight exit drills in 1958 was slightly more than three minutes. Had additional exit capacity been provided to accommodate the excessively high population density in the annex and south wing through erection of stair towers, slide escapes or fire escape stairs accessible from individual rooms the occupants of the second stories of these two sections could have reached the ground without having to pass through smoke-filled corridors and stairways.

As demonstrated by the loss of life in the north wing, however, adequate exit capacity is not the only consideration, or even the most important consideration, when evaluating the adequacy of exits. Of primary importance is the enclosure of exits to assure that the ways out of the building will be free of smoke and heat when needed.

Two Ways Out

Since there is always a possibility that fire or smoke may prevent the use of one exit, at least one alternate exit must be provided, remote from the first. Because of the fact that the three stairways from the second floor corridor of the north wing of Our Lady of the Angels School were all connected through the common corridor, the pupils in the second story classrooms had in reality no safe way out. The simple expediency of enclosing the three stairwells at the second story landings would have corrected this situation. The requirement for two exits could also have been met by erecting stair towers, slide escapes or fire escape stairs accessible from individual rooms.

Sprinklers and Exits

There is no question that if a complete, properly installed and adequately maintained automatic sprinkler system had been in Our Lady of the Angels School the fire at the base of the stairwell would have been quickly extinquished before smoke of any consequence had penetrated the upper story. It is, however, questionable practice to rely on fire extinguishment to the neglect of exits because of the possibility of both human and mechanical failure. Automatic sprinklers and stairway enclosures complement each other and both should have been installed in this school.

In existing buildings that lack enclosed exits and where it may be impractical or too expensive to enclose them, the Building Exits Code suggests the following substitute. The school building can be occupied safely if both 1) it is equipped with a standard automatic sprinkler system, and 2) if there is a standard exit of sufficient capacity from each room so that the occupants can escape without passing through any corridor which could be blocked by smoke, heat or fire. This condition may be met in various ways, such as providing doors leading directly outside from first floor rooms and by direct access to fire escape balconies from every room on upper floors.

Smoke Vents In Stairways

In the absence of an automatic smoke vent at the top of the stairwell where the fire started, all products of combustion from the fire in the stairwell were forced into the second story corridor. The presence of a vent would have reduced considerably the amount of smoke and hot fire gases that entered the corridor. A smoke vent, however, is not a substitute for proper stairway enclosures. It should be used in conjunction with such enclosures.

Interior Finish

It is generally recognized that in the interest of life safety in schools interior finish should be noncombustible (Class A, flame spread 0-20)[3] or at least slow burning (Class B, flame spread 20-75). The Building Exits Code, however, allows up to 10 per cent of the aggregate area of walls and ceilings of corridors and exitways to have a combustible Class C rating (flame spread 75-200). No interior finish with a higher flame spread rating is permitted. The wood trim in the second story corridor of the north wing with a Class C flame spread rating, represented about 17.5 per cent of the aggregate corridor area.

As previously indicated, there is difference of opinion as to whether or not the ceiling of the second story corridor was finished with combustible cellulose fiber acoustical tile. A finish of this material would increase by 23 per cent the aggregate corridor area with Class C (tile with flame retardant coating on exposed surface) or Class D (untreated tile). Because of the open stairways and the large amount of other combustible material present, the results of this fire can be satisfactorily explained without the presence of a combustible ceiling finish.

The ceilings of all classrooms in the second story of the north wing were finished with combustible cellulose fiber acoustical tile.

It is important to recognize the fact that even if this building had been of fire-resistive construction the results of this fire would have been similar because of the combustible material available at the bottom of the stairway, the absence of doors at the top of the stairway, and the combustible interior finish.

Detection

As again clearly demonstrated by this fire, the fact that a building is populated is no guarantee that a fire will be discovered promptly.

A complete automatic sprinkler system would have detected and extinguished the fire in its incipiency. An automatic fire detection system installed throughout the school would probably have discovered the fire before the second story corridor became impassable. Automatic protection in itself, however, is not a substitute for properly enclosed stairways. It is desirable supplementary protection.

In Case Of Fire

It is a cardinal rule of life safety that at the first indication of fire (which is usually smoke) all occupants of the building and the fire department should be alerted simultaneously. From the time the teacher of Room 206 was first told that there was smoke in the building until she operated the building fire alarm, it is estimated that 13 minutes elapsed. Her actions during this vital period have already been described under "Discovery." They clearly indicate that adequate steps had not been taken at this school to assure proper emergency action by the teachers in case of fire, nor had a sufficient number of building fire alarm controls been provided.

Had the building fire alarm been rung when the fire was discovered it is probable that the second story corridor of the north wing would still have been passable.

The alarm system at this school was not connected to the fire department alarm headquarters, and no one in the school telephoned the fire department.

Fire Alarms

Interior Alarms

The substandard condition of the manual fire alarm system in the school should be noted. The alarm sending switches were not readily accessible to most of the occupants of the building, they were only two in number and were not distinguishable from ordinary electric light switches. Had readily identifiable fire alarm stations been distributed throughout the building it is likely that at least one of the people who first noticed smoke would have operated the building alarm system many minutes sooner.

Exterior Alarms

According to the present Municipal Code of Chicago and the recognized standard for the installation of public fire alarm boxes[4], all schools should have a fire alarm box at or near the entrance. Had a box been so located at Our Lady of the Angels School it is probable that the first alarm would have been transmitted to the fire department by the passerby at least one or two minutes earlier. Furthermore, the box alarm would have resulted in response of a box alarm assignment on the first notification of the fire. This would have brought three additional engine companies and one additional ladder company to the scene three or four critical minutes earlier.

The NFPA Building Exits Code recommends that any building fire alarm system be arranged so that when operated to alert occupants of the building it will simultaneously transmit an alarm to the fire department. Automatic sprinkler systems and automatic detection systems should be arranged to operate building and fire department alarm systems simultaneously.

Housekeeping

Following the fire the remains of a large amount of combustible material (bundled newspapers, exam papers, etc.) was found among the debris at the base of the stairwell where the fire started. The school authorities stated that combustible material was not supposed to be accumulated in this area. At the base of and under each of the two front stairwells, however, there was a wooden storage closet in which wooden chairs, screen panels and other combustible materials were stored; and a former pupil of the school stated that in 1957 newspapers from a paper drive were stored at the foot of the rear stairway. Good housekeeping is thus again emphasized as a cardinal fire safety principle.

Summary

The ninety-three deaths in this fire are an indictment of those in authority who have failed to recognize their life safety obligations in housing children in structures which are "fire traps." Schools that lack adequate exit facilities and approved types of automatic sprinkler or detection equipment, and which possess excessive amounts of highly combustible interior finish, substandard fire alerting means and poor housekeeping conditions must be rated as "fire traps." School and fire authorities must take affirmative actions to rid their communities of such blights.

---

[1] The nationally recognized NFPA Building Exits code is the standard reference on adequate exit facilities from Buildings.

[2] After reviewing this report Fire Commissioner R.J. Quinn stated that he disagreed with the conclusion reached by the NFPA investigator, that the principal cause of loss of life was inadequate exit facilities. Commissioner Quinn, feeling that the exits in this building were adequate, believes that the principal cause of loss of life was delayed alarm to the fire department.

[3] Interior finish materials are classified by the tunnel test method (NFPA No. 255) in which asbestos-cement board has a rating of 0 on the flame spread scale and red oak lumber a rating of 100.

[4] NFPA No.73, standard for installation, maintenance and use of Municipal Fire Alarm Systems.

I have belabored this report and statistics because, sadly, I must also report, just in this short period since the completion of my investigation, numerous accounts of other disastrous fires have appeared in the daily newspapers.

Mankind never seems to learn that fire is an inexorable foe lurking always to destroy man's property and to consume mankind itself. Unfortunately, man does not have a fire-proof skin.

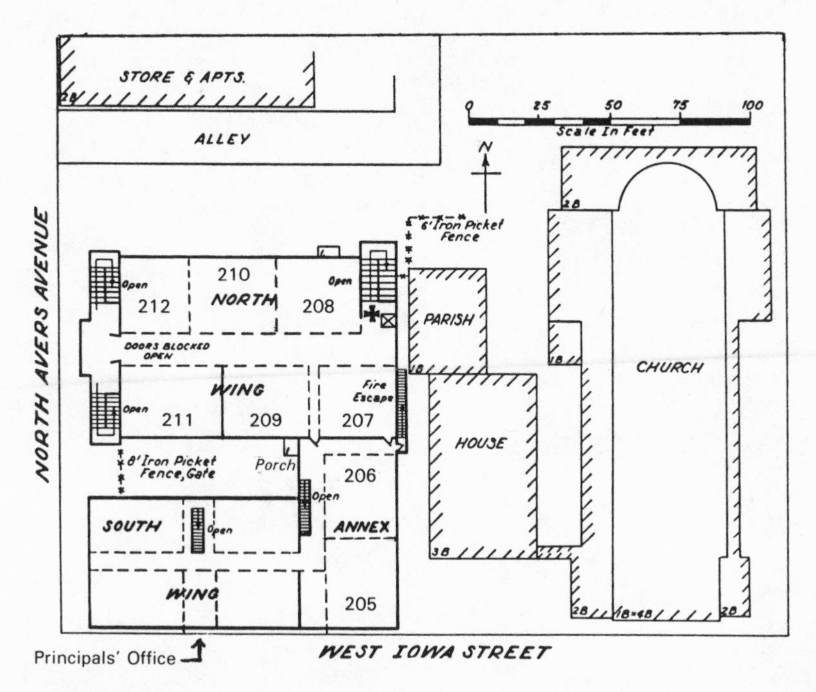

Diagram 1. - Plot plan of school and adjacent property. Second story details of school building shown. Fire started at basement level of stairway marked by cross.

Death Toll As Of August, 1959

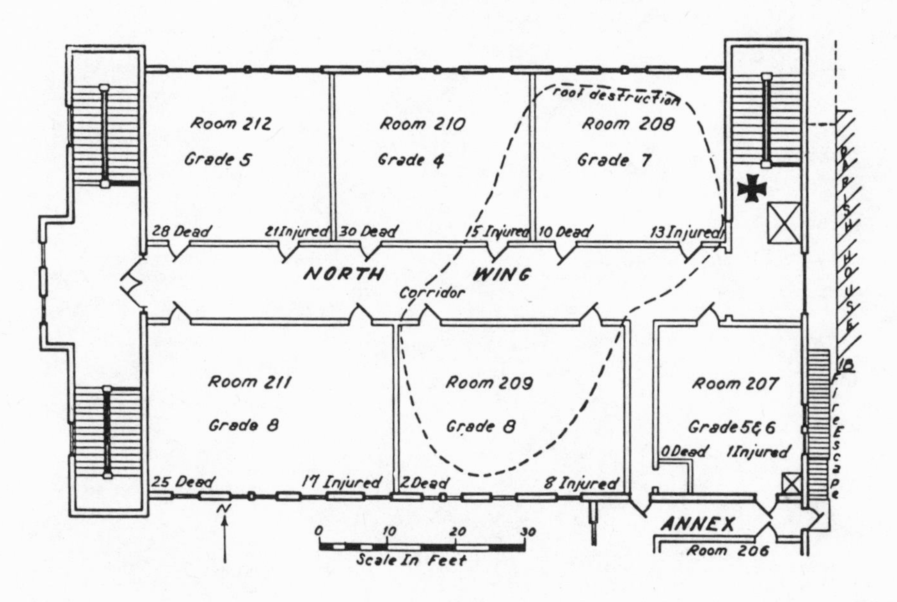

Diagram 2. - Second story of north wing. Fire started at basement level of stairway marked by cross.

Michele McBride on her Easter, 1959, visit to her home with her dog Daisy. (Chicago Tribune Photo)

The view of the school which faced the alley. Note the short ladders which were first put in place by the firemen.
(World Wide Photo)

Second floor corridor that was filled with fire trapping the students in the classrooms. The ceiling caved in from the intense heat. (World Wide Photo)

Charred stairwell where it is suspected the fire started. (World Wide Photo)

Remains of the classroom where twenty-eight perished. (World Wide Photo)

The funeral mass at the Armory for twenty-four children who died in the Our Lady of the Angels fire. (Chicago Tribune Photo)

Our Lady of the Angels Monument in Queen of Heaven Cemetery, Hillside, Illinois, dedicated to those who lost their lives in Our Lady of the Angels fire. (Photo by Michele McBride)